604.25 NE
Nelson, John A., 1935-
How to read and understand
blueprints /

604.25 NE
Nelson, John A., 1935-
How to read and understand
blueprints /

D1202012

South St. Paul Public Library
106 Third Avenue North
South St. Paul, MN 55075

HOW TO READ
AND UNDERSTAND
BLUEPRINTS

HOW TO READ
AND UNDERSTAND
BLUEPRINTS

John A. Nelson

VNR VAN NOSTRAND REINHOLD COMPANY
NEW YORK CINCINNATI TORONTO LONDON MELBOURNE

Copyright © 1982 by Van Nostrand Reinhold Company Inc.

Library of Congress Catalog Card Number: 81-15917
ISBN: 0-442-26188-8

All rights reserved. No part of this work covered by the copyright hereon may
be reproduced or used in any form or by any means—graphic, electronic, or
mechanical, including photocopying, recording, taping, or information storage
and retrieval systems—without permission of the publisher.

Manufactured in the United States of America

Published by Van Nostrand Reinhold Company Inc.
135 West 50th Street, New York, N.Y. 10020

Van Nostrand Reinhold Limited
1410 Birchmount Road
Scarborough, Ontario M1P 2E7, Canada

Van Nostrand Reinhold Australia Pty. Ltd.
17 Queen Street
Mitcham, Victoria 3132, Australia

Van Nostrand Reinhold Company Limited
Molly Millars Lane
Wokingham, Berkshire, England

15 14 13 12 11 10 9 8 7 6 5 4 3 2 1

Library of Congress Cataloging in Publication Data

Nelson, John August, 1935-
    How to read and understand blueprints.

    Includes index.
    1. Blue-prints.   I. Title.
T379.N37   604.2'5   81-15917
ISBN 0-442-26188-8   AACR2

Q
604.2
Ne
C.1

To my daughter, Joy

South St. Paul Public Library
106 Third Avenue North
South St. Paul, MN 55075

PREFACE

Today everyone should know how to read and understand blueprints. For employment requirements, promotion opportunities, or simply self-satisfaction, knowing how to interpret todays drawings is important to almost everyone. This text is designed for both self-study and/or class use and is presented in a logical, step-by-step sequence with practice exercises emphasizing the material just learned.

The content presented is slanted somewhat toward the mechanical field of drafting used in industry but most of the "practices" are used in reading and understanding architectural drawings, also. The material in this text is presented one of four ways:

1. *"Lecture"* pages, which simply explain, one or more basic drafting practices.
2. *"Worksheet"* pages, which ask questions directly related to the material that has just been covered in the lecture pages. Most worksheets include working with drawings, similar to those used in industry, which give the reader an opportunity to apply the material just studied and learned to an actual drawing.
3. *"Terms to know"* pages, are important terms the reader should know. These terms are actually the "language of the trade." It is suggested that the reader write out a brief definition of each term in the space provided as each term is explained on the lecture pages.
4. At the end of each chapter there is a *"Final Evaluation"* which covers all or most of the drafting practices just learned. It is important that the reader fully understand all terms and questions on the final evaluation before going on to the next chapter.

Answers for all tests and question pages are fully illustrated, in order, at the end of each chapter. It is recommended that readers check each answer as they proceed through the text.

ACKNOWLEDGMENTS

The author wishes to thank David Leavitt, Teacher, St. Johnsbury, Vermont and David Ziller, Manuscript Editor at Van Nostrand Reinhold Company for reviewing the manuscript and providing critical input.

The instructional material in this text was classroom tested in the vocational department of Contoocook Valley Regional High School, Peterborough, New Hampshire.

Special thanks to my daughter Joy for testing this material.

CONTENTS

HOW TO READ
AND UNDERSTAND
BLUEPRINTS

UNIT 1

THE TITLE BLOCK

Objective: To learn what a standard title block should contain, what various symbols mean, what size paper is used in industry, and how to calculate limits and tolerances.

TERMS TO KNOW AND UNDERSTAND

As you read and study, use this page for *notes.* See end of chapter for answers.

Paper Sizes _____

_____

Folding Paper _____

_____

Title Block (what it contains) _____

_____

_____

_____

Tolerancing Of Dimensions _____

_____

_____

Revision (Change) Block _____

_____

_____

_____

PAPER SIZES

Drawing paper, tracing paper, vellum, linen, or drawing plastic films are available in sheets or rolls. Standard sizes are noted in the chart. Roll stock usually comes in rolls of 34 inches (841 mm) x 50 yards long and is designated as "R" size.

Paper is "called off" as "A" size, "B" size etc., which designates a particular size. Notice the large space 1 1/4 inch on the left hand size of the paper. This is to provide space for *binding* pages together.

PAPER SIZE			
Size	Fraction (inch)	Metric (mm)	Size
A	8 1/2 x 11	210 x 297	A-4
B	11 x 17	297 x 420	A-3
C	17 x 22	420 x 594	A-2
D	22 x 34	594 x 841	A-1
E	34 x 44	841 x 1184	A-0

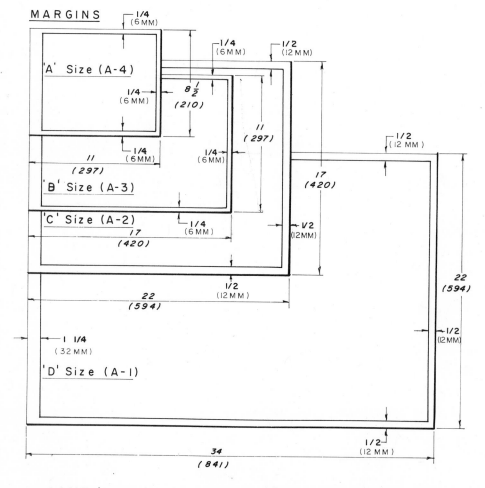

A.S.M.E. (American Society of Mechanical Engineers) standard margin size.

FOLDING PAPER

All paper, regardless of its size usually is filed in a standard file and thus, must be folded. Each size is a multiple of 8 1/2 x 11 (210 x 297), and when folded as illustrated below, reduces to an "A" size (A-4) or *8 1/2 x 11* (210 x 297).

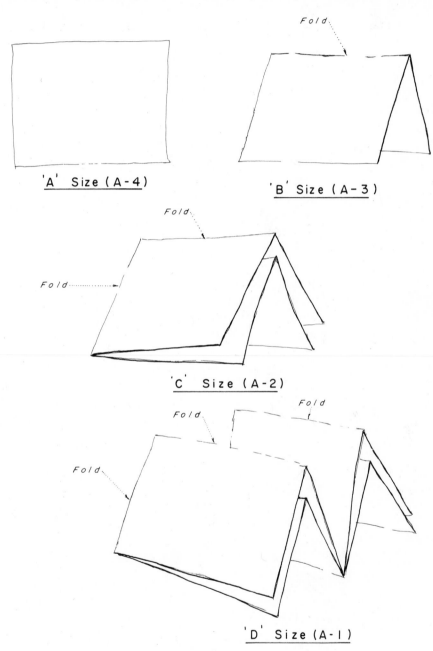

'A' Size (A-4)   'B' Size (A-3)

'C' Size (A-2)

'D' Size (A-1)

TITLE BLOCK

The title block of a drawing is usually placed in the lower right hand corner of the drawing. It is usually printed on the paper along with the standard and is filled in, upon completion of the drawing, freehand, using uppercase (capital) vertical letters by the draftsperson.

There are many styles of title blocks but most usually contain the following information:

1. Company name
2. Company address
3. Drawing title
4. Drawing number
5. Scale(s) used
6. Tolerancing of all dimensions
7. Material(s) used to make the part
8. Heat-treatment process (if any)
9. Finish requirements (if any)
10. Draftsperson's name and date drawn
11. Checker's name and date checked
12. Person approving drawing for company and date approved
13. Revision (change) record block
14. Projection symbol (1st or 3rd angle)
15. Dimension notation (inch/metric)
16. Page number (if more than one page)

Title block items are listed below in more detail:

1. *Company name*—self-explanatory

2. *Company address*—self-explanatory

3. *Drawing title*—For example, "bracket-hinge." Note the dash between the words. In 'calling-off' the name of the part, it is referred to as a *"hinge bracket"* (*not* bracket-hinge) there are various parts of the hinge but *this* particular part is the *bracket,* thus, called the hinge bracket.

4. *Drawing number*—Companies use many various numbering systems, some a ten digit numbering system to make use of the computer for record keeping. Because there are various paper sizes used in an engineering department. Each size has a letter designation, A, B, C, or D, in order to note the paper size, many companies use a prefix of A, B, C, D, (paper-size) *before* the actual drawing number. Example B-77532, the *B* indicates the drawing is done on a "B" size sheet of paper 11 x 17 inches. A drawing, or part of the drawing, is changed from time to time and the change is recorded in the "revision block" (to be explained later) in order to call attention to the *last* change, some companies add the last revision (change) letter at the *end* of the part number, example: B-77532-C. The "C" at the end of the number indicates, "C" change was the last revision.

5. *Scale(s) used*—The 'scale' or proportion, to actual size, the drawing is drawn to. A very large part must be "scaled" down to fit on a sheet of paper. The part could be drawn 1/2 size, 1/4 size or even 1/10 size. A very *small* part could be drawn 2x (twice size) or even 10x (10 times size) regardless of the scale used, the *dimensions* indicate the exact size the part is to be manufactured.

6. *Tolerancing of dimensions*—(To be covered in detail later.) Indicates the permissable variation of size allowed from the actual indicated dimension.

7. *Material*—The exact material(s) that is to be used to manufacture the part.

8. *Heat-treatment process*—Indicates the exact specifications that must be followed to harden (or soften) the part.

9. *Finish requirements*—This indicates if the part is to be primed, painted, chrome plated, etc. (Many times, there is simply a *note* above the title block).

10. *Draftsperson's name/date*—The person who drew the drawing signs and dated this part.

11. *Checker's name/date*—The person who checked the drawing signs and dated this part.

12. *Approval*—The person who approves the part and thus, releases the drawing for production.

13. *Revision–(change) record block*–It is extremely important that an accurate systematic record be kept of all changes made *since the drawing was approved*. It is in this area that all change information is recorded, consecutively from A to Z. In the event a major change is made, the drawing is completely redrawn. The original drawing is filed in the obsolete file and the *next* change letter added to the new drawing.

B	WAS 6¹/16	2/10/79
A	ADDED NOTE	12/11/78
LET	CHANGE	DATE

CHANGE NOTICE

14. *Projection symbol*–There are two methods used to graphically illustrate a part. The United States and Canada uses *third-angle projection* while the rest of the world uses *first-angle projection*. Due to the increase in international exchange of parts and drawings the method used should be indicated on the drawing. In the event this is not indicated simply note the origin or the country the drawing came from. Third-angle projection will be illustrated at this time. Later in the text, *first-angle projection* will be illustrated. The symbols used to denote which system is being used is illustrated:

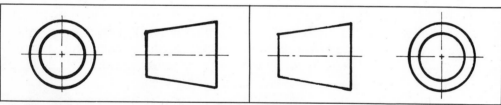

THIRD ANGLE PROJECTION   FIRST ANGLE PROJECTION

15. *Dimension notations*–With the advent of the metric system a translation period will exist for many years, thus, a *dual* dimensioning system, (inch system *and* the metric system) of dimensioning will be indicated. A notation such as this will be noted in the title block:

$$\frac{inch}{(Millimeter)}$$

This means all dimensions are in inches with its *equivalent* in metric in parenthesis ( ). For example:

$$\frac{5}{(127)}$$

The upper figure is the controlling dimension (the other, simple the converted value).

16. *Page number*–Used if there is more than 1 page and indicated, for example, as "page 2 of 5" meaning this is page 2 of 5 complete pages.

All company title blocks differ but should contain most of the material noted on this sample title block. Referring to the circled numbers, study each of the 16 items listed below. Be sure you understand what each is, why it is included and how it should be used.

1. Company name
2. Company address
3. Drawing title (note the *dash* between words)
4. Drawing number (note the prefix "B" which indicates a B-size paper and the suffix "C" which indicates the *last* change was "C"
5. Scale used
6. Tolerancing of all dimensions
7. Material used
8. Heat-treatment process
9. Finish requirements (many times contained in a "note" above the title block)
10. Draftsperson's name/date *completed*
11. Checker's name/date checked
12. Persons name approving drawing/date approved
13. Revision (change) record block
14. Projection symbol
15. Dimension notation
16. Page number

Notice the note *"Do Not Scale Drawing."* A craftsperson should *never* scale, or measure, to find a missing dimension.

NOTE: SPRAY PAINT w/ #7762-5 ENAMEL

DRAWN JAN 6/19/78	CHECKED SRP 6/20/78	INCH MILL
MATERIAL CAST IRON	APPROV. C.J.M. 6/30/78	PAGE 1 OF 1
	TITLE BRACKET - HINGE	
	JAN ENGINEERING MAIN ST. ST. JOHNSBURY, VT.	DRAWING NO. B-7753Z-C

DO NOT SCALE DRAWING		
TOLERANCE UNLESS OTHERWISE SPEC.	.XX ± .015 .XXX ± .010 .XXXX ± .0002	
HEAT TREATMENT NONE		
SCALE 1/2 SIZE		

LET	CHANGE	DATE
C	ADDED NOTE	10/18/79
B	ADDED 1/4 DIA. R.	8/6/79
A	WAS 2 1/16 LONG	10/5/78
	CHANGE NOTICE	

TOLERANCE AND LIMITS

Think of the signs along interstate highways that tell how fast or slow you are allowed to travel. If you go faster or slower than posted you could be stopped and fined.

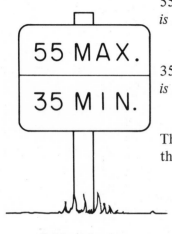

55 M.P.H. is the fastest you are allowed to travel. *This is the 'high limit.'*

35 M.P.H. is the slowest you are allowed to travel. *This is the 'low limit.'*

The *tolerance* in this example is the difference between the high limit and the low limit thus:

$$
\begin{array}{r}
55 \text{ mph (high limit)} \\
-35 \text{ mph (low limit)} \\
\hline
= 20 \text{ mph accepted } tolerance
\end{array}
$$

Driving tolerances

Tolerance is the difference between the limits.

For example, a drawing must state the largest and the smallest size that is acceptable for a particular application or function. Thus, in this example:

.505 is the largest hole allowed (high limit).
.500 is the smallest hole allowed (low limit).

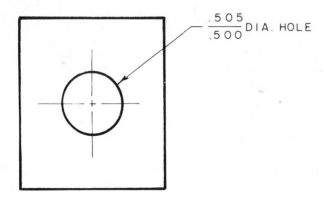

The *tolerance* in this example is the difference between the high limit and the low limit. Thus:

$$
\begin{array}{r}
.505 \text{ diameters (high limit)} \\
- .500 \text{ diameters (low limit)} \\
\hline
= .005 \text{ Accepted } tolerance
\end{array}
$$

The draftsperson should *never* ask for closer or tighter limits than are necessary. Closer or tighter limits require better machinery, better trained craftsman, and much more time to make and thus, *will cost more.*

TOLERANCING OF DIMENSIONS

Nothing can be manufactured *exactly* to the correct size as specified on a drawing. All sizes end up either larger or smaller than the dimension called for. Thus, a method to allow for this variation must be followed:

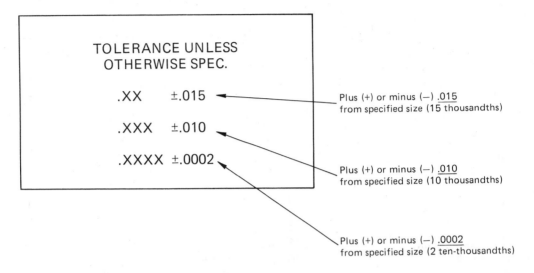

Using the chart:

1. Any fraction dimensions are considered not critical and can vary up to *plus* (+) 1/64 (2 mm) or *minus* (−) 1/64 (2 mm) from the specified size called off on the drawing.
2. Any two-place (.xx) decimal dimensions are also not considered critical thus, they too can vary up to *plus* (+) .015 (15 thousandths) or *minus* (−) .015 (15 thousandths) from the specified size called off on the drawing.
3. A three-place (.xxx) decimal dimension would be considered more important or more critical, thus, the allowed variable from the *required size* would be held much closer. Up from specified size only, plus (+) .010 (10 thousandths) or down minus (−) .010 (10 thousandths) from the specified size.
4. A four-place (.xxxx) decimal dimension would be considered very critical and a variance from the required size would be only +.0002 (2 ten-thousandths) or −.0002 (2 ten-thousandths).
5. Any dimension that must be held to a specific or *exact* size would be dimensioned to the exact, allowed size, such as .250/.251 dia. which means the hole cannot be any *smaller* than .250 dia. and cannot be any larger than .251 dia. in size.

It is important that the person reading the drawing note the tolerance block before proceeding too far. Each company will have their own tolerance requirements.

TOLERANCING WORKSHEET

Given below is specified size *"dimension"* (left hand column) and the "tolerance unless otherwise specified" block. Using these two columns, calculate the upper/lower limits and the tolerance of each problem below. See end of chapter for answers.

DIMENSION	TOLERANCE UNLESS OTHERWISE SPEC.	UPPER LIMIT	LOWER LIMIT	TOLERANCE
1/2	NOT NOTED (1/64 Standard)			
.50	.XX ± .015			
.500	.XXX ± .010			
.5000	.XXXX ± .0002			
.250/.251	AS 'CALLED-OFF'			
3.875	.XXX ± .005			
1.0032	.XXXX ± .0001			
3/4	NOT NOTED			

TITLE BLOCK WORKSHEET I

Answer each question. See end of chapter for answers.

1. What is the *title* of this drawing?

2. What is the *number* of this drawing?

3. In the drawing number, what does the prefix "A" mean?

4. What is the last change letter and when was it changed?

5. What date was this drawing checked?

6. What page number is this drawing?

7. What *scale* is used in this drawing?

8. What was done to this drawing August, 1981?

9. A .xxx placed decimal has a *tolerance* of what?

10. What is the company name?

INCH / MILL		
CHECKED JAN 7/18/79	7/22/79	PAGE 2 OF 5
DRAWN SRP 7/18/79	APPROV. RCB 7/30/79	TITLE SUPPORT-SHELF
MATERIAL ALUMINUM		DRAWING NO. A-8293-D
		NELSON BROS. R.F.D. 4 ST. JOHNSBURY, VT.

LET	CHANGE	DATE
DO NOT SCALE DRAWING		
TOLERANCE UNLESS OTHERWISE SPEC. XX ± .015 XXX ± .005 XXXX ± .00Ci		
HEAT TREATMENT NONE		
SCALE FULLSIZE		
D	WAS 5 1/64 LG.	10/9/81
C	ADDED NOTE	8/2/81
B	CHANGED MAT'L	7/3/81
A	ADDED FINISH	1/12/80
LET	CHANGE	DATE
CHANGE NOTICE		

TITLE BLOCK WORKSHEET II

Answer each question. See end of chapter for answers.

1. Is this drawing done in 1st angle projection or 3rd angle projection? How can you tell?

2. What is the drawing number?

3. What size paper is used? How is it designated?

4. What scale is used?

5. What date was this drawing approved? By whom?

6. What finish is used on this part?

7. What change was made and when?

8. What does the inch/Mill symbol mean?

9. What is the angular tolerance (unless otherwise noted)?

10. What is the part made of?

11. The dimension 3.750 would have limits of what? Tolerance of?

12. What are the upper/lower limits for a dimension of 15.50?

MODERN INDUSTRY
MAIN ST.
ST. JOHNSBURY, VERMONT

LINK-DRAG
(MAIN LINKAGE ASSEMBLY)

DRAWING NO.
352-5514-A

SIZE
A

SCALE FULL

SHEET 1 OF 1

	DATE
DR. COPPAGE	2/4/79
CH. NELSON	2/10/79
APP. EDWARDS	2/12/79

INCH
(MILL)

HEAT TREATMENT
NONE

UNLESS OTHERWISE SPECIFIED:
DIMENSIONS ARE IN INCHES
TOLERANCES: FRACTIONS ± 1/64
ANGLES ± 0°-30'
3 PLACE DECIMALS ± .005
2 PLACE DECIMALS ± .015

MATERIAL
1010 STEEL

FINISH
CHROME PLATE

EVALUATION

See end of chapter for answers.

1. All drawing sizes, when folded, are reduced to what dimensions?
2. What size paper is a 'B' size (A-3)?
3. Why is there an 1 1/4 inch margin on the left hand side of 'standard' A.S.M.E. paper?
4. What is the metric size for a "D" size drawing? How is it called off?
5. What kind of lettering is found on most drawing title blocks?
6. How would you call off a part that had the title—*"Screw-Adjusting"*?
7. A drawing has a number *C-734410-A*, what does the "C" *indicate and what does the* "A" *indicate?*
8. In the space labeled *"Scale"*, there is a *"2x"*, what does this mean?
9. What does this symbol mean? What countries uses this symbol?

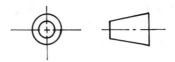

10. What would $\dfrac{inch}{(Millimeter)}$ indicate?
11. Why do many companies add the note: *"Do Not Scale Drawing"*?
12. A dimension is called off as: .750/.760, what is the tolerance?
13. A dimension of *8.521* +/−.005 would have limits of what?
14. The *"Tolerance unless otherwise spec."* block indicates .xx = +/−0.15. What would the limits and tolerance be for .75?
15. Which would cost the most to manufacture, a hole that was *.50 dia.* or one *.500 dia.?* Why?

ANSWERS TO TERMS TO KNOW AND UNDERSTAND

```
Paper Sizes — A =  8 1/2 x 11  –  210 x  297  –  A – 4
              B =    11 x 17   –  297 x  420  –  A – 3
              C =    17 x 22   –  420 x  594  –  A – 2
              D =    22 x 34   –  594 x  841  –  A – 1
              E =    34 x 44   –  841 x 1184  –  A – 0
```

Folding Paper—All paper regardless of size, folds down to an 'A' size (A-4) 8 1/2 x 11.

Title Block Contains—Company name/company address/title/number/scale/tolerancing of all dimensions/material/heat-treat process/finish requirements/draftsperson's name/checker's name/approval signature/revision block/projection symbol/dimension notation and page number.

Tolerancing Of Dimensions—The permissional variation of size allowed from the actual indicated dimension.

Revision (Change) Block—A systematic record of any and all changes made since the drawing was approved.

Symbol—3rd angle projection/used by U.S. and Canada.

TOLERANCING WORKSHEET ANSWERS

Given below is specified size *'dimension'* (left hand column) and the "tolerance unless otherwise spec." block. Using these two columns, calculate the upper/lower limits and the tolerance of each problem below. (Place answer in the space provided.)

DIMENSION	TOLERANCE UNLESS OTHERWISE SPEC.	UPPER LIMIT	LOWER LIMIT	TOLERANCE
1/2	NOT NOTED...... 1/64 Standard	33/64	31/64	1/32
.50	.XX ± .015	.515	.485	.030
.500	.XXX ± .010	.510	.490	.020
.5000	.XXXX ± .0002	.5002	.4998	.0004
.250/.251	AS CALLED-OFF	.251	.250	.001
3.875	.XXX ± .005	3.880	3.870	.010
1.0032	.XXXX ± .0001	1.0033	1.0031	.0002
3/4	NOT NOTED	49/64	47/64	1/32

TITLE BLOCK WORKSHEET I ANSWERS

1. *Shelf* support (support–shelf would be *incorrect*).
2. A-8293-D.
3. The drawing is an *A size* (A-4) paper size or 8 1/2 x 11 (210 x 297).
4. "D" was the last change, it was changed 10/9/81.
5. The drawing was checked 7/22/79.
6. This is page 2 of 5 pages.
7. The drawing was made *"full size."*
8. A 'note' was added 8/2/81.
9. .xxx = $\pm$ .005.
10. The company name is *"Nelson Bros."*

TITLE BLOCK WORKSHEET II ANSWERS

1. 3rd angle projection is used on this drawing. The *symbol* in the title block illustrates 3rd angle.
2. A-352 – 5514-A (The first *A* indicates drawing size.)
3. 8 1/2 x 11 inch size paper—it is designated by an "A."
4. Scale = full size.
5. 2/12/79 approval was by Edwards.
6. Chrome plate
7. 1 3/8 was changed to 1.391, 10/17/79.
8. The inch/(Mill) symbol indicates all dimensions are in inches with the metric equivilant in parentheses underneath.
9. +/– 0° - 30′
10. 1010 steel
11. +/–.005 limits/.010 tolerance
12. 15.50 dimension = 15.515 upper limit/15.485 lower limit. (It is a two-placed decimal thus, plus or minus .015)

EVAULATION ANSWERS

1. 8 1/2 x 11/'A' size (210 x 297/A-4).
2. 'B' size (A-3) = 11 x 17 (297 x 420).
3. The 1 1/4 inch margin at the left is to allow space for binding pages together.
4. A 'D' size drawing is 594 x 841 mm and is called off as A-1.
5. Upper case (capital) vertical letters, done freehand by the draftsperson.
6. The correct call off would be *"adjusting screw."*
7. The *"C"* indicates the drawing is a 'C' size; the *'A'* indicates the *last* revision or change was 'A.'
8. 2x means twice size; the drawing was drawn twice its actual size.
9. Means 3rd angle projection. Only the United States and Canada uses this system.
10. Means the drawing is in *inches* with *millimeter* equivalent in parentheses ( ) under the inch. Example:

$$\frac{1\ 1/2}{(38.1)}$$

11. A craftsperson should *never* scale or measure to find a missing dimension.
12. .750/.760 = .010 tolerance. (The difference between the limits).
13. *8.526* (upper limits)/*8.516* (lower limits)
    (8.521 + .005 = 8.526/8.521 − .005 = 8.516)
14. *.765* (upper limits)/.735 (lower limit)/.030 tolerance
    (.750 + .015 = .765/.750 − .015 = .735)
15. The hole *.500 dia.* would cost the most because the size tolerance is smaller.

UNIT 2

CONVERTING TO METRIC

Objective: To practice converting inch to millimeters and millimeters to inches and study how this is applied to the drawing.

TERMS TO KNOW AND UNDERSTAND

See end of chapter for answers.

Rounding Off _____
_____
_____

Controlling Dimension _____
_____
_____
_____

Converting _____
_____
_____
_____

Design Size _____
_____
_____
_____

Millimeter = Inch _____
_____
_____

Inch = Millimeter _____
_____
_____

CONVERTING

For years to come, the United States, Canada, and the rest of the world will go through a transitional period from the old inch system to a metric system. Until that time, everyone will have to convert from one system to the other. This can be done mathematically or by using conversion charts. An international system has been initiated, to provide a logical method to accomplish this transition.

A conversion chart has been provided or you may simply use the factor 1 mm = 0.03937 inch/1 inch = 25.400 mm.

ROUNDING OFF DIMENSIONS

If you are mathematically rounding off, calculate the figure to one greater number of digits than required and round off by one of the following methods:

1. Increase the last required digit by one, if the digit which follows, to the right, *is more than 5.*
2. Leave the last required digit the same, if the digit which follows, to the right, *is less than 5.*

Example:

5.7216   Round off to 3 decimal places = 5.722
6.2534   Round off to 3 decimal places = 6.253
1.673   Round off to 2 decimal places = 1.67
0.777   Round off to 2 decimal places = 0.78

CONTROLLING DIMENSIONS

The title block should contain a *dimension notation.* This will be illustrated by either an *inch/(millimeters)/*symbol or millimeter/(inch) symbol. This actually is double dimensioning, with the important *controlling dimension* on top. The *top* figure would usually be the system the country, where the drawing was drawn, uses as their standard.

Study the illustration below, which is an example of what each system will look like with its double-dimension or converted equivalent underneath. As noted in title block . . .

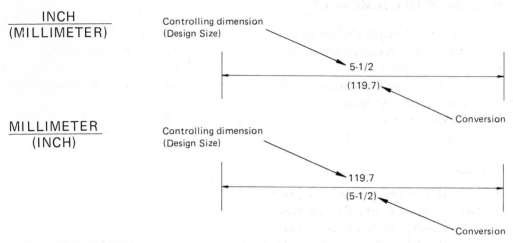

INCH
(MILLIMETER)

Controlling dimension
(Design Size)

5-1/2

(119.7)

Conversion

MILLIMETER
(INCH)

Controlling dimension
(Design Size)

119.7

(5-1/2)

Conversion

CONVERSION CHART

There are many kinds of conversion charts available. Illustrated below is a simple chart which converts from inch fractions to inch decimals to millimeters. If you know *any* of the three you can simply read off the other two.

FRACTION	DECIMAL	METRIC
1/64	.016	.395
1/32	.031	.790
3/64	.047	1.190
1/16	.062	1.585
5/64	.078	1.985
3/32	.094	2.380
7/64	.109	2.775
1/8	.125	3.175
9/64	.1	3.573
5/32		

Example: 3/64 inch = .047 inch = 1.190 mm

Example: .078 inch = 5/64 inch = 1.985 mm

Example: 3.175 mm = .125 inch = 1/8 inch

CONVERSION CHART

INCH/METRIC – EQUIVALENTS					
Fraction	**Decimal Equivalent**		**Fraction**	**Decimal Equivalent**	
	Customary (in.)	**Metric (mm)**		**Customary (in.)**	**Metric (mm)**
1/64 —.015625		0.3969	33/64 —.515625		13.0969
1/32 —.03125		0.7938	17/32 —.53125		13.4938
3/64 —.046875		1.1906	35/64 —.546875		13.8906
1/16 —.0625		1.5875	9/16 —.5625		14.2875
5/64 —.078125		1.9844	37/64 —.578125		14.6844
3/32 —.09375		2.3813	19/32 —.59375		15.0813
7/64 —.109375		2.7781	39/64 —.609375		15.4781
1/8 —.1250		3.1750	5/8 —.6250		15.8750
9/64 —.140625		3.5719	41/64 —.640625		16.2719
5/32 —.15625		3.9688	21/32 —.65625		16.6688
11/64 —.171875		4.3656	43/64 —.671875		17.0656
3/16 —.1875		4.7625	11/16 —.6875		17.4625
13/64 —.203125		5.1594	45/64 —.703125		17.8594
7/32 —.21875		5.5563	23/32 —.71875		18.2563
15/64 —.234375		5.9531	47/64 —.734375		18.6531
1/4 —.250		6.3500	3/4 —.750		19.0500
17/64 —.265625		6.7469	49/64 —.765625		19.4469
9/32 —.28125		7.1438	25/32 —.78125		19.8438
19/64 —.296875		7.5406	51/64 —.796875		20.2406
5/16 —.3125		7.9375	13/16 —.8125		20.6375
21/64 —.328125		8.3384	53/64 —.828125		21.0344
11/32 —.34375		8.7313	27/32 —.84375		21.4313
23/64 —.359375		9.1281	55/64 —.859375		21.8281
3/8 —.3750		9.5250	7/8 —.8750		22.2250
25/64 —.390625		9.9219	57/64 —.890625		22.6219
13/32 —.40625		10.3188	29/32 —.90625		23.0188
27/64 —.421875		10.7156	59/64 —.921875		23.4156
7/16 —.4375		11.1125	15/16 —.9375		23.8125
29/64 —.453125		11.5094	61/64 —.953125		24.2094
15/32 —.46875		11.9063	31/32 —.96875		24.6063
31/64 —.484375		12.3031	63/64 —.984375		25.0031
1/2 —.500		12.7000	1 —1.000		25.4000

CONVERTING WORKSHEET I

Using the "given" dimension below, and the conversion chart, list the converted answer in the space provided. See end of chapter for answers.

| PROB. | INCH | | METRIC mm |
	FRACTION	DECIMAL	
1	1/4		
2		.625	
3			24.606
4	1 3/8		
5		2.734	
6			50.80
7	3 1/2		
8		2.750	
9			47.625
10		6875	

CONVERTING WORKSHEET II

Using the conversion chart, carefully figure what the millimeter equivalents will be for each dimension below. Fill in answers in parentheses, ( ), under each decimal inch dimension. (There are eight answers). See end of chapter for answers.

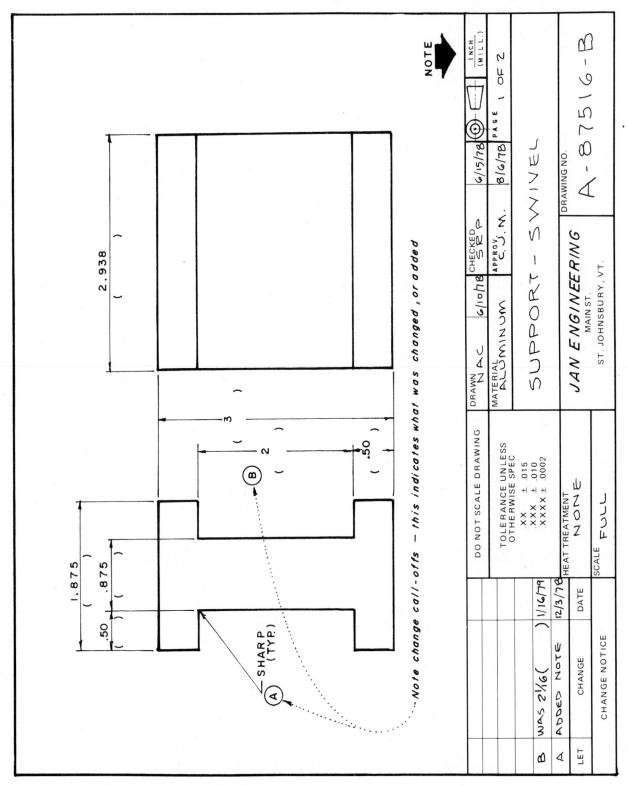

REVIEW

Using the previous title block, answer the following questions about the title block. See end of chapter for answers.

1. What is the *title* of the part?

2. What is the drawing number?

3. What does the prefix *"A"* mean?

4. What does the suffix *"B"* indicate?

5. When was the drawing checked?

6. What scale is used on this drawing?

7. What change was made 12/3/78?

8. What *was* the 2 (50.80) dimension when the drawing was first issued?

9. What is the material?

10. What are the upper and lower *limits* on the 2.938 (74.62) dimension?

11. What is the *tolerance* of the .50 (12.70) dimension?

12. What does the inch/(Mill.) symbol in the title block indicate?

EVALUATION

Instructions: Fill in all ten converted equivalences in areas provided. See end of chapter for answers.

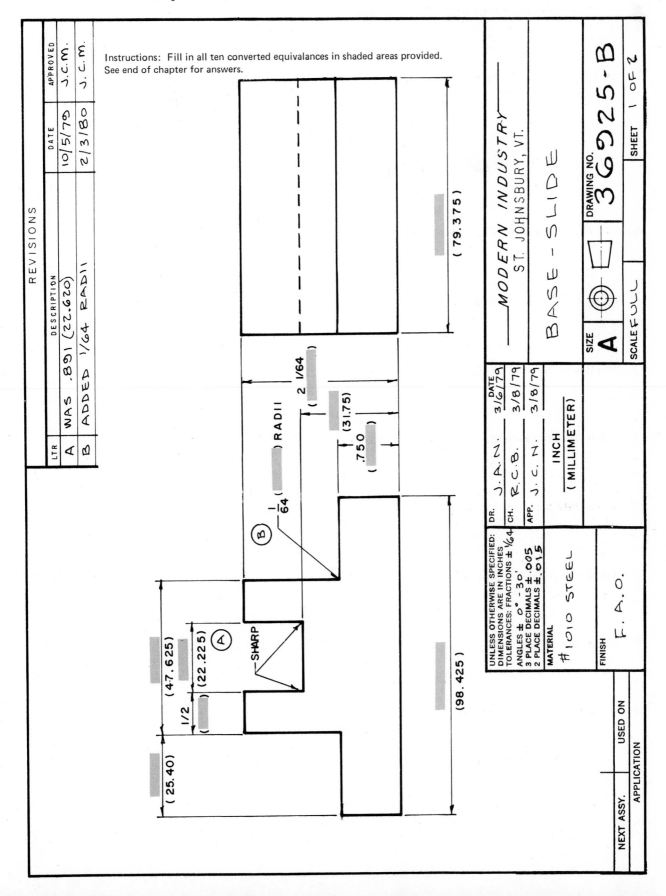

EVALUATION

Answer each question. See end of chapter for answers.

1. Explain what the *controlling dimension* is.

2. Where on the drawing does it indicate the controlling dimension system used? How is this done?

3. What is the fraction equivalent and millimeter equivalent for 0.938 decimal?

Refer to the drawing on p. 00 for the following:

4. What is the drawing number?

5. What size paper is it on?

6. What material is used to make this part?

7. What date was this part approved?

8. What was done on the last change? (B)

9. What are the upper/lower *limits* of the 1/2 inch (12.70) dimension?

10. What is the *tolerance* for the 1/64 (.395) radii?

ANSWERS TO TERMS TO KNOW AND UNDERSTAND

Rounding off—Is the process of calculating a figure one greater number of digits than required and increasing the last required digit by one of the digits to the right if more than 5, or leaving the last required digit the same, if the digit to the right, is less than 5.

Controlling dimension—When the double dimensioning system is employed; i.e., inch/metric, the figure on *top* is the *controlling dimension.*

Converting—Changing a figure in one measuring system to another system by either calculation or conversion charts.

Design size—Actually the *controlling dimension.* This is the system that was used to design the part (either inch or metric).

Millimeter = inch—1.0 mm = 0.394 inch

Inch = millimeter—1.0 inch = 25.4 mm

CONVERTING WORKSHEET I ANSWERS

PROB.	INCH FRACTION	INCH DECIMAL	METRIC mm
1	1/4	.250	6.350
2	5/8	.625	15.875
3	31/32	.969	(24.605) 24.606
4	1 3/8	1.375	25.400 + 9.525 / 34.925 34.925
5	2 47/64	2.734	69.450
6	2	50.80 -25.40 / 25.40 2.00	50.80
7	3 1/2	3.50	25.4 × 3 / 76.2 +12.7 / 88.9 88.90
8	2 3/4	2.750	25.40 +25.40 +19.05 / 69.85 69.85
9	1 7/8	47.625 -25.400 / 22.225 1.875	47.625
10	11/16	.6875	17.470

CONVERTING WORKSHEET II ANSWERS

Notice the inch/(mm) symbol in the title block.
This indicates the *inch system* is the *controlling dimension.*

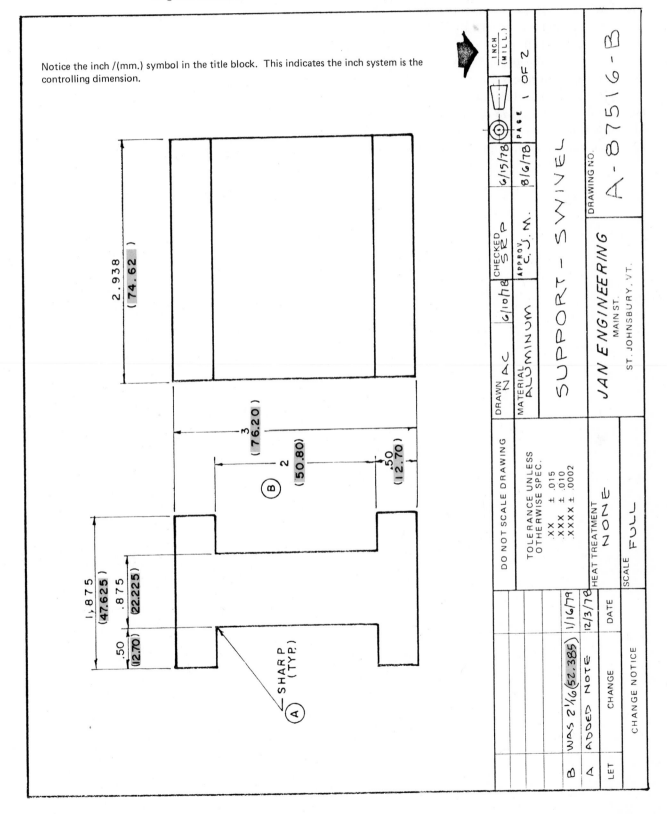

Notice the inch /(mm.) symbol in the title block. This indicates the inch system is the
controlling dimension.

REVIEW ANSWERS

1. The parts title is *Swivel Support.*
2. Drawing *#A-87516-B.*
3. The "A" means, *'A' size paper* (8 1/2 x 11).
4. The "B" indicates the last *change* was "B" (1/16/79).
5. The drawing was checked 6/15/78.
6. The scale is full size.
7. The note, *"Sharp-Typ."* was added.
8. The dimension was *2 1/16 inch (52.385),* when first issued.
9. The material is aluminum.
10. 2.938 + 010 = *2.948 upper limit/*2.938 – .010 = *2.928 lower limit.*
11. .XX tolerance = +/– .015. or *.030*
12. Inch/(Mill.) symbol indicates the *inch* system is the controlling dimensions.

EVALUATION ANSWERS

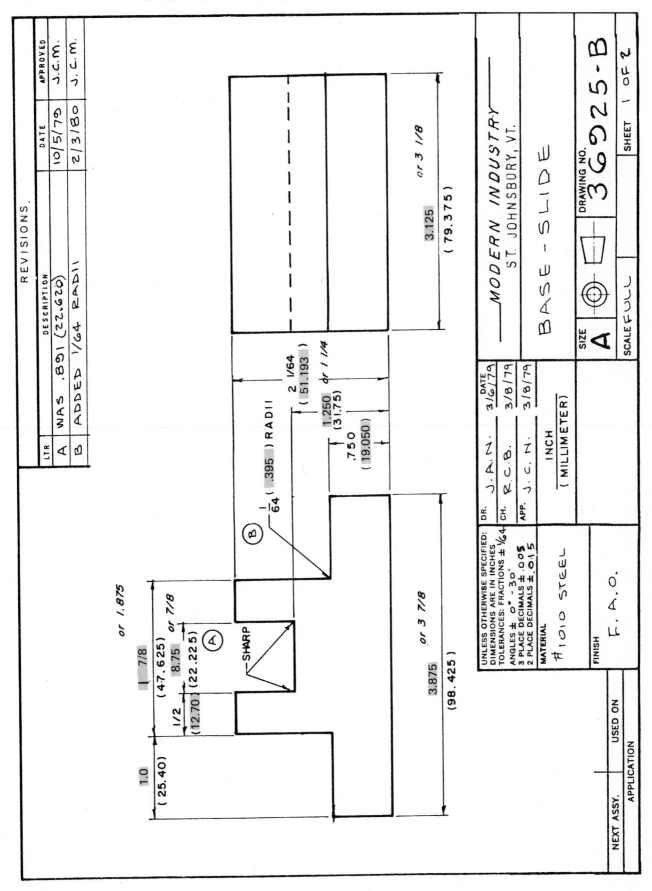

EVALUATION ANSWERS

1. The *controlling dimension* is the figure (dimension) on *top*. The equivalent dimension is noted below it. The figure on *top* indicates the system used to design the part.
2. The system of dimensions used should be indicated in the title block. It is called off by either inch/(millimeter) or millimeter/(inch).
3. .938 = 15/16 inch (23.820 mm).
4. The drawing number is A-36925-B.
5. The paper size is "A" or 8 1/2 x 11.
6. The material is #1010 steel.
7. It was approved 3/8/79.
8. A 1/64 radii was added.
9. 1/2 + 1/64 = 33/64 *upper limit*
   1/2 − 1/64 = 31/64 *lower limit*
10. All fractions = *1/32* (+/− 1/64).

UNIT 3

KINDS OF LINES

Objective: To be able to identify various kinds of lines used in industry to illustrate and dimension an object.

TERMS TO KNOW AND UNDERSTAND

As you study try to memorize each *kind* of line, its *thickness,* how it is represented and *what* it represents. (Sketch each if it will help). See end of chapter for answers.

Object Line_____

Hidden Line_____

Center Line_____

Extension Line_____

Dimension Line_____

Leader_____

Cutting Plane Line_____

Section Line_____

Phantom Line_____

THICKNESS OF LINES

A drawing is made up of many lines. Each line represents *something:* A surface, a hidden surface, a center of a hole, an extension of a surface, or a line with dimensions on it. In order to make the drawing easier to read and understand, each kind of line is drawn with a different thickness. There are three thickness of lines: thick, medium thick, and thin. Study this illustration, note the different thicknesses of all three lines.

NOTE THICKNESS
(THICK)

NOTE THICKNESS
(MEDIUM)

NOTE THICKNESS
(THIN)

THICK LINES INCLUDE:

or

MEDIUM LINES INCLUDE:

THIN LINES INCLUDE:

x.xx

(xx·xx)

Object Line—Illustrates all visible *edges* of the object drawn.

Cutting Plane Line—Shows where a section has been taken. Arrows indicate the direction section was taken.

Hidden Line—Shows *surfaces* that cannot be seen. This is a dash line.

Center Line—Locates center of parts, holes or surfaces, drawn by long and short dashes.

Extension Line—Extends surface that is to be dimensioned.

Dimension Line—Shows extent of dimension also include dimension and arrows.

Section Line—Illustrated surface that has been cut.

Phantom Line—Shows position(s) of part of an object that moves drawn by two short dashes and one long dash.

KINDS OF LINES

The shape of an object is drawn by *object lines.* They are solid heavy lines.

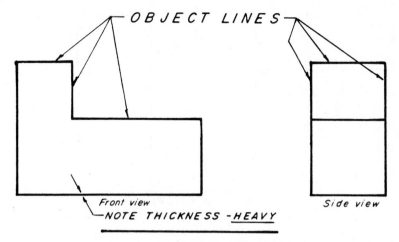

Any surface that cannot be seen, such as the side view of the two holes below, are drawn as *hidden lines.* They are short dashed, medium thick lines.

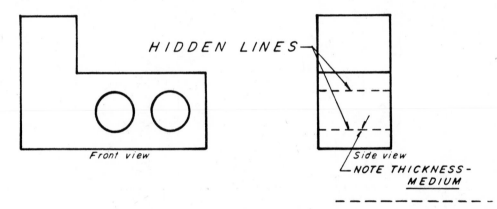

Center lines indicate the middle (center) of an object, a circle, or the center of a radius. This is *not* part of the object and thus is drawn very thin.

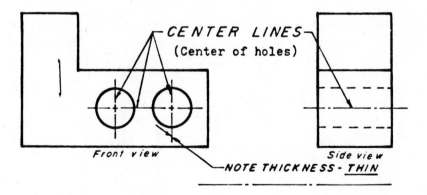

Extension lines as the name implies, *'extends'* from the object. They are thin lines.

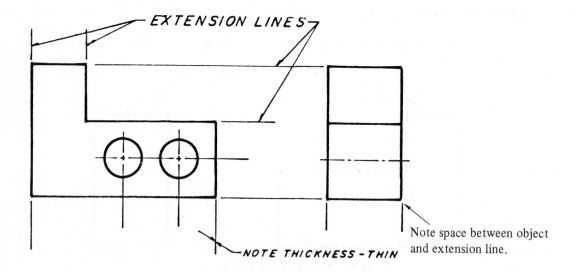

Dimension lines show the extent of a dimension and include the dimension line and arrowheads. They are *thin* lines.

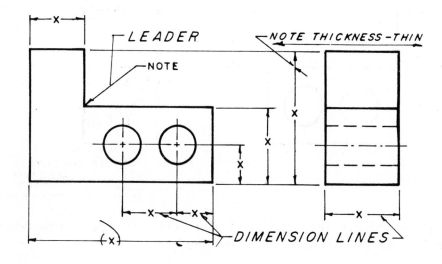

Section lines illustrate the *surface* that has been cut. They are thin lines.

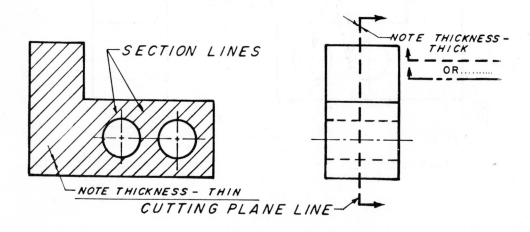

Cutting plane lines show where a section has been taken. They all are either short dashes or two short dashes to one long dash.

Study the drawing below, it contains many kinds of lines.

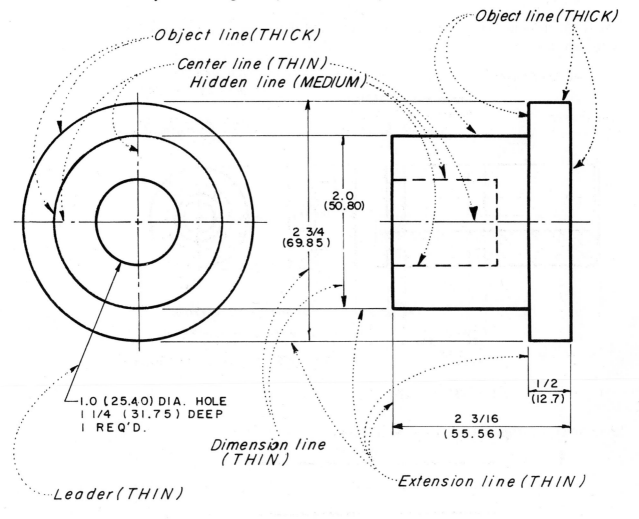

As you review, note each kind of line in the above drawing:

Object line—Thick lines illustrate all visible edges and surfaces.

Hidden line—Medium lines, dash line; shows hidden surfaces. The hidden lines above show where the 1 inch (25.40) dia. hole is located.

Center line—Thin, long and short dash line; shows the center of the parts *and* center of the 1 inch (25.40) dia. hole.

Extension line—Thin, extends from the object in order to add dimension line.

Dimension line—Thin, shows extent of dimensions, also includes dimensions and arrows.

Leader—Thin, exactly like a dimension line except usually has a *note* at one end as illustrated.

Study the drawing below, note the two new kinds of lines, section lines and cutting plane line.

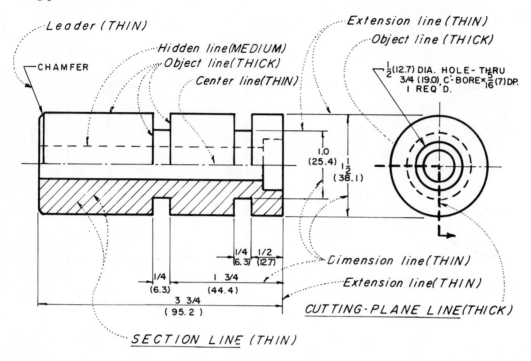

Remember, a cutting plane line can be drawn as

➖ ➖ ➖ ➖ ➖ ➖ ➖ ➖

or

➖ ➖ ➖➖ ➖ ➖➖ ➖ ➖➖

As you review, note each kind of line in the above drawing.

Cutting plane line—Thick line, shows where a section has been taken (this will be covered later). Arrows indicate the *direction* the section was taken.

Section line—Thin line, illustrates surface that has been cut.

Phantom line—Thin line, shows position(s) of part of an object that moves–drawn by two short dashes and one long dash, (think of the seewaw).

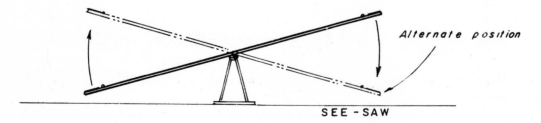

SEE - SAW

KINDS OF LINES WORKSHEET

Using the previous pages as review, study the various kinds of lines used to draw this object. List, in the spaces provided below, the kinds of lines labeled *A* through *N*. See end of chapter for answers.

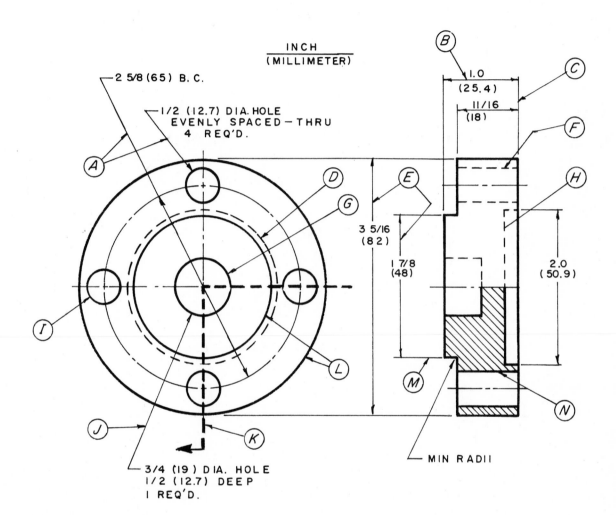

South St. Paul Public Library
106 Third Avenue North
South St. Paul, MN 55075

Answers

A _____ B _____ C _____ D _____

E _____ F _____ G _____ H _____

I _____ J _____ K _____ L _____

M _____ N _____

KINDS OF LINES EVALUATION

1. Using drawing A–794376 (page 42) identify lines *A* through *P* in the spaces below. See end of chapter for answers.

A _____ B _____ C _____ D _____

E _____ F _____ G _____ H _____

I _____ J _____ K _____ L _____

M _____ N _____ O _____ P _____

2. What is the metric equivalent of dimension "Q"?

3. What is the 'Title' of the drawing?

4. How many changes have been made on this part to date?

5. What is the material?

6. How many pages are included for this drawing?

7. When was the drawing approved?

8. What is the tolerance for the 1.50 dimension?

9. How many thicknesses of lines are there?

10. What are the upper/lower limits for the 1.875 dimension?

EVALUATION

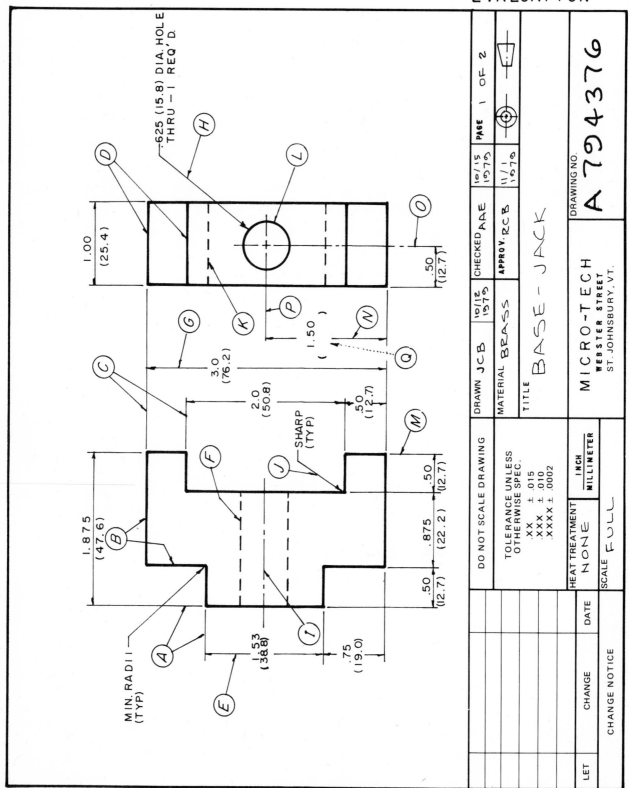

DO NOT SCALE DRAWING	DRAWN JCB	10/12 1976	CHECKED AAE	10/15 1976	PAGE 1 OF 2
	MATERIAL BRASS		APPROV. RCB	11/1 1976	
	TITLE BASE - JACK				
TOLERANCE UNLESS OTHERWISE SPEC. XX ± .015 .XXX ± .010 .XXXX ± .0002	MICRO-TECH WEBSTER STREET ST. JOHNSBURY, VT.			DRAWING NO. A 794376	
HEAT TREATMENT NONE	INCH / MILLIMETER				
SCALE FULL					
LET	CHANGE	DATE			
CHANGE NOTICE					

-.625 (15.8) DIA. HOLE THRU - 1 REQ'D.

1.00 (25.4)

3.0 (76.2)

2.0 (50.8)

.50 (12.7)

.50 (12.7)

1.50

.50 (12.7)

.875 (22.2)

.50 (12.7)

1.875 (47.6)

1.53 (38.8)

.75 (19.0)

SHARP (TYP)

MIN. RADII (TYP)

ANSWERS TO TERMS TO KNOW AND UNDERSTAND

Object Line–Thick, illustrates all visible edges or surfaces of the object drawn.

Hidden Line–Medium, shows surfaces that cannot be seen (short dashes).

Center Line–Thin, locates the center of parts, holes or surfaces. (Long and short dashes).

Extension Line–Thin, extends surface that is to be dimensioned.

Dimension Line–Thin, shows extent (or length) of dimensions–includes dimension and arrows.

Leader–Thin, similar to a dimension line but usually has a *note* at one end.

Cutting Plane Line–Thick, shows where a section has been taken–has *either* short dashes *or* 2 short dashes/long dash–has arrows indicating direction section was taken.

or

Section Line–Thin, illustrates the surface that has been cut.

/////

Phantom Line–Thin, shows position(s) of part of an object that moves (drawn by two short dashes and one long dash).

KINDS OF LINES WORKSHEET ANSWERS

A Leader	*B* Dimension	*C* Extension	*D* Hidden
E Dimension	*F* Hidden	*G* Object	*H* Hidden
I Object	*J* Leader	*K* Cutting plane	*L* Object
M Extension	*N* Object		

KINDS OF LINES EVALUATION ANSWERS

1. *A* Extension   *B* Object   *C* Extension   *D* Object

    E Dimension   *F* Hidden   *G* Dimension   *H* Leader

    I Center   *J* Leader   *K* Hidden   *L* Object

    M Extension   *N* Dimension   *O* Center   *P* Center

2. 1.50 inch = 38.1 mm.
3. Base-jack (called off as *"jack-base"*)
4. *No* changes to date
5. Material = *Brass*
6. Two pages, this is page *1* of 2
7. Drawing was approved 11/1/79
8. .XX = (+/− .015), thus .030 tolerance
9. Three: thick, medium, and thin
10. Upper limit = 1.875 + .010 = 1.885
    Lower limit = 1.875 − .010 = 1.865

UNIT 4

MULTIVIEW DRAWINGS

Objective: To read and understand one, two, and three view drawings similar to those used in industry.

TERMS TO KNOW AND UNDERSTAND

Aligned Dimensioning System _____

Unidirectional Dimensioning System _____

Out-Of-Scale Dimensions _____

Metric Diametral Dimension Symbol _____

One View Drawing _____

B.C. (Bolt Circle) _____

Two View Drawings _____

Projection Lines _____

Three View Drawings _____

Width, Height, and Depth _____

Kinds Of Holes Used In Industry _____

Boss Or Pad _____

DIMENSIONING SYSTEMS

There are two methods to place dimensions on the drawing. The aligned system is not used much any more.

Aligned

All dimensions are *aligned* with the dimension line and read from either the bottom or right side of the paper.

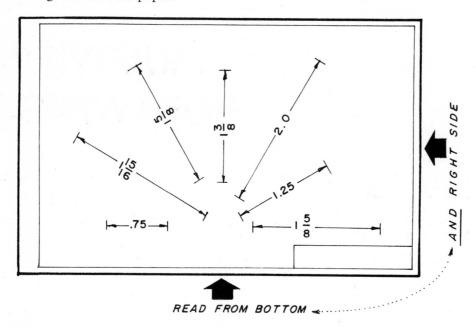

Unidirectional

In the unidirectional system, all dimensions are read from the bottom of the page as illustrated. This is the newer of the two systems.

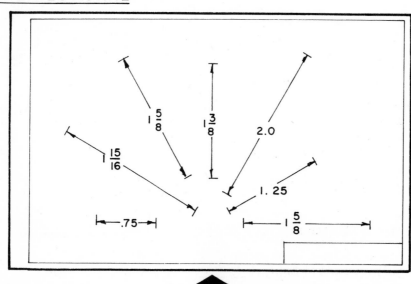

OUT OF SCALE DIMENSIONS

If any part of a drawing is not to scale, a wavy line is placed *under* the dimension.

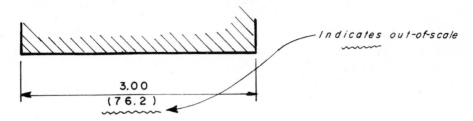

DIMENSIONS

A drawing must represent the object or part *exactly* as it is to be manufactured and must convey each and every detail without question. The drawing expresses the required sizes and locations by a series of dimensions which could be either fractions, decimals or millimeters—or a combination of two or more. Remember, dimension lines are thin, solid lines which terminate with arrowheads. The required size or location dimension is the *distance between the arrowhead points*. Note each example below (In A, the extension lines are used in conjunction with the dimension lines. In D, the required location and size is between the arrowhead points):

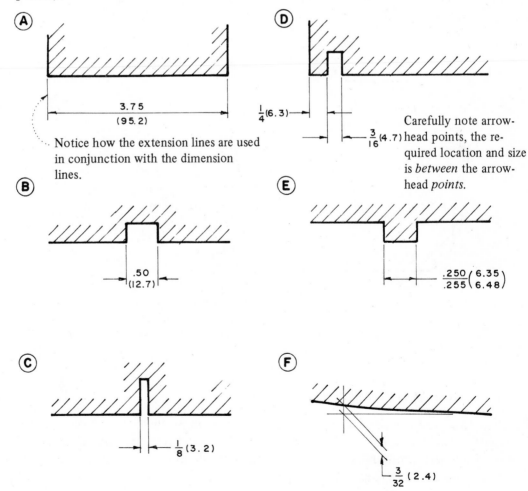

DIMENSION AND RADII

A circular arc is dimensioned by giving its radius. If there is enough space, the radius dimension line is drawn from the radius center, with an arrowhead ending on the circular arc. (See A and B). The "R" indicates *radius*. Many times there is not enough space for the dimension, thus the dimension may be placed outside the dimension line as illustrated (If the swing-point falls off the paper, the drafts-person illustrates the dimension with a "break" as in E):

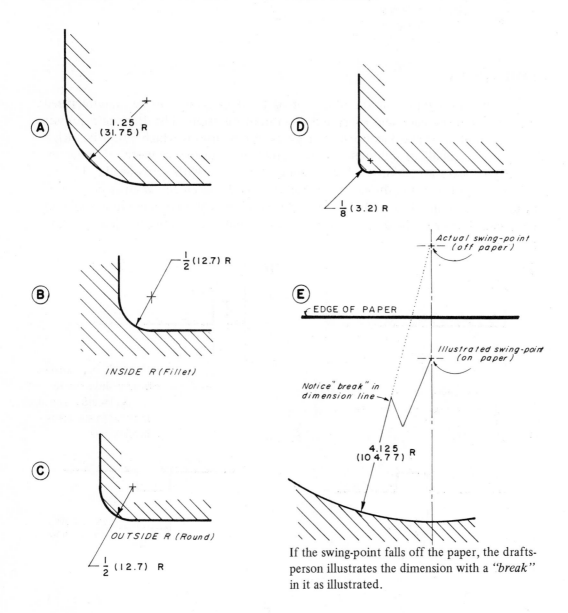

If the swing-point falls off the paper, the drafts-person illustrates the dimension with a *"break"* in it as illustrated.

DIMENSION AND DIAMETERS

Diameters, whether they are holes or round areas on an object should be dimensioned as illustrated below. The center of the diameter is located by the intersection of the two center lines. On *all* holes, the note should include diameter, depth, and number of holes required.

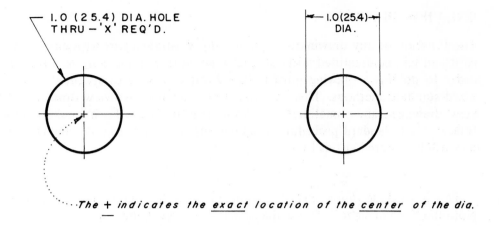

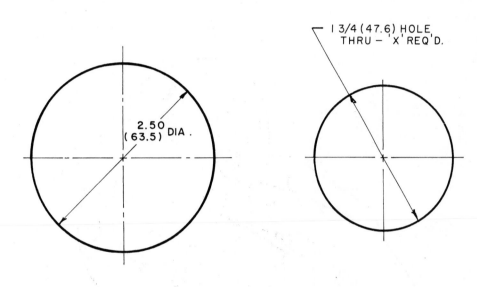

METRIC DIAMETRAL DIMENSION SYMBOL: φ

A drawing using the metric system as its controlling, or design system will use this symbol in the place of the word *Dia.*

Example: 25.4 (1.0) φ

ONE-VIEW DRAWINGS

The function of any drawing is to graphically illustrate a part with enough details so it can be manufactured without question and turn out *exactly* as intended. In order to do this a drawing must be layed out according to accepted and recognized standard methods. There are one-view drawings, two-view drawings, three-view drawings, and sometimes even four-view drawings. Each kind of drawing follows a set standard procedure in laying out the views. Simple, one-view drawings will be covered at this time.

This is a simple *thin* gasket. If you held it in your hand it would look like this. Note the "line of sight." This is the direction you view the object.

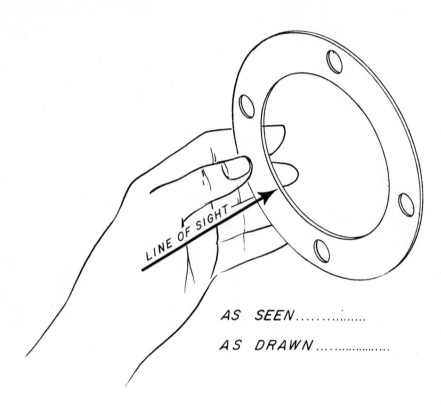

LINE OF SIGHT

AS SEEN..............

AS DRAWN...............

This same object would be illustrated like this. You are looking *directly* at the *front* of the gasket. (Its *thickness* is called-off in a note.)

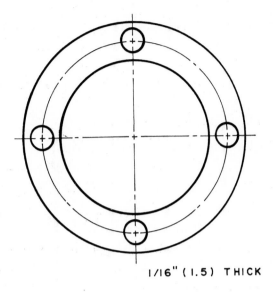

1/16" (1.5) THICK

ONE VIEW DRAWINGS WORKSHEET I

Using drawing A621117-C, answer the following questions. Because the *inch* system is the "design size," use the inch system for all answers. See end of chapter for answers.

1. Assuming the center line indicates the center of the link, what is dimension 'A'? _____

   _____

2. Figure dimension 'B'—figure answer in decimals to three places._____

   _____

3. What is the metric equivalent of 'C'? _____

4. Identify each kind of line as indicated by:
   D_____   E _____   F _____
   G_____   H _____

5. What does 'I' indicate? _____

   _____

6. What *was* the overall length when the drawing was issued? _____

   _____

7. What *scale* is the drawing drawn to? _____

8. How thick is the part? _____

9. How many .50 (12.7) dia. holes are there? _____

10. Which would cost the most—the .50 dia. hole(s) or the .625 dia. hole? Why?

    _____

    _____

11. What is the center to center distance of the .625 (15.875) dia. hole and the 1.00 (25.4) dia. hole? _____

    _____

12. What is the distance between the centers of the two .50 (12.7) dia. holes?

    _____

    _____

ONE VIEW DRAWING

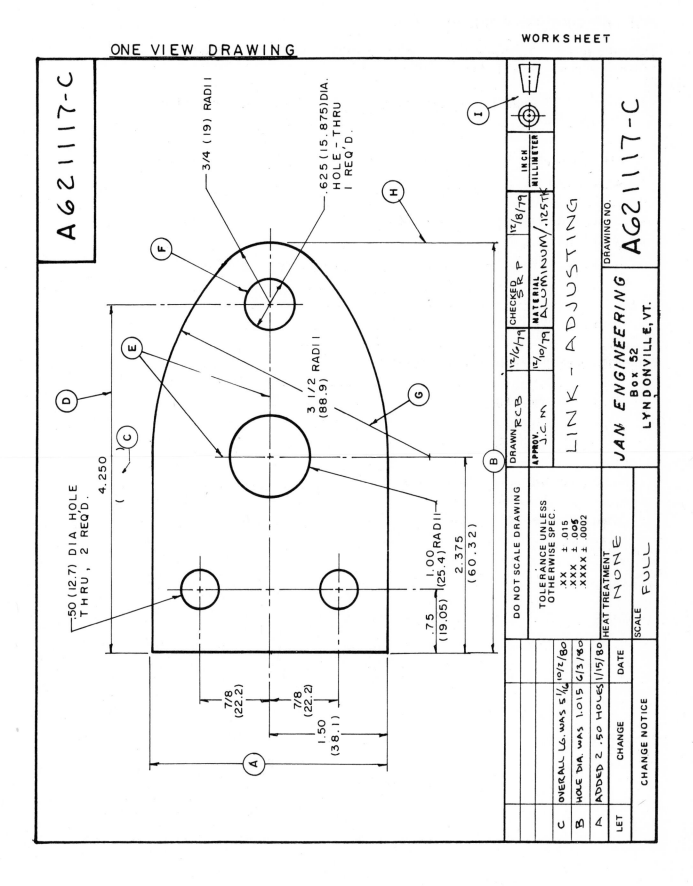

A621117-C

3/4 (19) RADII

.625 (15.875)DIA.
HOLE - THRU
1 REQ'D.

3 1/2 RADII
(88.9)

.50 (12.7) DIA HOLE
THRU, 2 REQ'D.

4.250

1.00
(25.4) RADII

.75
(19.05)

2.375
(60.32)

7/8
(22.2)

7/8
(22.2)

1.50
(38.1)

DRAWN RCB   12/6/79   CHECKED SRP   12/8/79   INCH MILLIMETER
APPROV. J.C.M   12/10/79   MATERIAL ALUMINUM/.125TK

LINK - ADJUSTING

JAN ENGINEERING
Box 52
LYNDONVILLE, VT.

DRAWING NO. A621117-C

DO NOT SCALE DRAWING

TOLERANCE UNLESS
OTHERWISE SPEC.
.XX ± .015
.XXX ± .005
.XXXX ± .0002

HEAT TREATMENT NONE

SCALE FULL

C   OVERALL LG. WAS 5 1/16   10/2/80
B   HOLE DIA. WAS 1.015   6/3/80
A   ADDED 2 .50 HOLES 1/15/80

LET   CHANGE   DATE

CHANGE NOTICE

ONE VIEW DRAWING WORKSHEET II

Using drawing A4937, answer the following questions. (Answers in inches unless otherwise noted). See end of chapter for answers.

1. Assuming the center line indicates the center of the shim, what is dimension 'A'? _____
   _____

2. Figure dimension 'B' _____ inches (_____) MM.

3. What is radii 'C'? _____
   _____

4. Calculate dimension 'D' in millimeters _____ MM.

5. What is the *decimal* answer of 'E'? _____

6. Assuming the center line indicates the center of the shim, what is dimension 'F'? _____
   _____

7. What kind of a line is 'G'? _____
   _____

8. What does the notation at "H" mean? _____
   _____

9. What is the spacing between the .50 (12.7) dia. holes? _____
   _____

10. How many .50 dia. holes are there? _____
    _____

11. What is the tolerance on all *fractions?* _____
    _____

12. What is the material made of? _____
    _____

ONE VIEW DRAWING

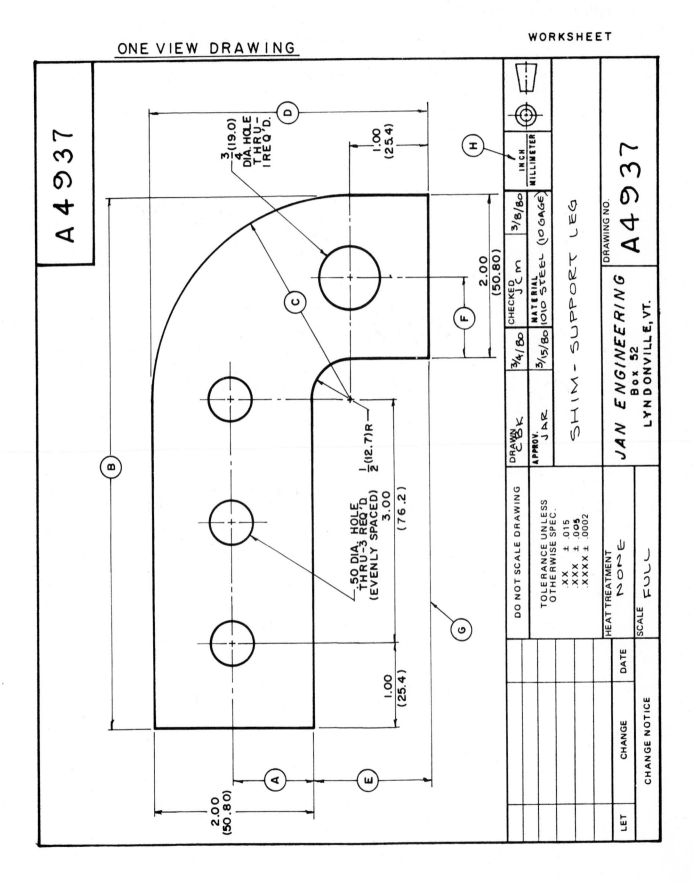

A4937

D

1.00
(25.4)

3/4 (19.0)
DIA. HOLE THRU-
1 REQ'D.

C

B

1/2 (12.7) R

.50 DIA. HOLE
THRU-3 REQ'D
(EVENLY SPACED)

3.00
(76.2)

1.00
(25.4)

2.00
(50.80)

2.00
(50.80)

F

G

H

A

E

INCH
MILLIMETER

| DRAWN
CBK | 3/4/80 | CHECKED
JCM | 3/8/80 |
| APPROV.
JAR | 3/15/80 | MATERIAL
1010 STEEL (10 GAGE) | |

SHIM-SUPPORT LEG

JAN ENGINEERING
Box 52
LYNDONVILLE, VT.

DRAWING NO.
A4937

DO NOT SCALE DRAWING

TOLERANCE UNLESS
OTHERWISE SPEC.

.XX ± .015
.XXX ± .005
.XXXX ± .0002

HEAT TREATMENT
NONE

SCALE FULL

DATE

CHANGE

LET

CHANGE NOTICE

DIMENSION/LOCATION HOLES

There are various methods used to locate holes on a circle or diameter. Study these illustrations. Note how the example on the left actually is the same as on the right except *called-off* in note below. B.C. means *Bolt Circle* (The diameter the holes are located on).

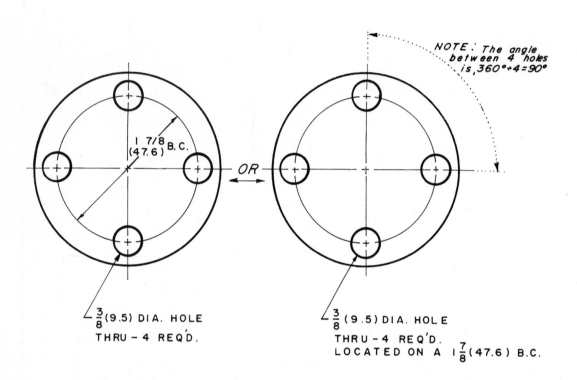

NOTE: The angle between 4 holes is, 360°÷4=90°

1 7/8 (47.6) B.C.

OR

$\frac{3}{8}$ (9.5) DIA. HOLE
THRU – 4 REQ'D.

$\frac{3}{8}$ (9.5) DIA. HOLE
THRU – 4 REQ'D.
LOCATED ON A 1$\frac{7}{8}$ (47.6) B.C.

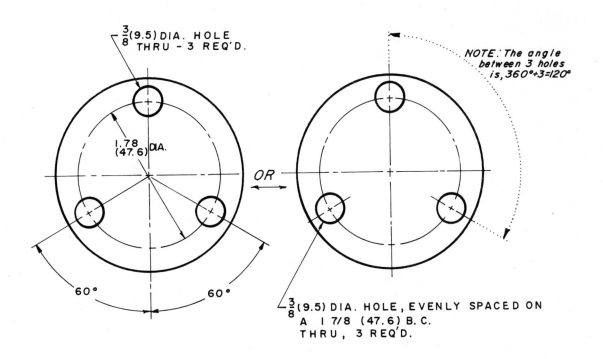

$\frac{3}{8}$ (9.5) DIA. HOLE
THRU - 3 REQ'D.

1.78
(47.6) DIA.

NOTE: The angle
between 3 holes
is, 360°÷3=120°

OR

60°   60°

$\frac{3}{8}$ (9.5) DIA. HOLE, EVENLY SPACED ON
A 1 7/8 (47.6) B. C.
THRU, 3 REQ'D.

Review: B.C.= Bolt Circle (Diameter size)

ONE VIEW DRAWINGS WORKSHEET III

Using drawing A63279-A, answer the following questions. (Remember there are 360° in a circle). See end of chapter for answers.

1. What is the O.D. (outside diameter) of the gasket? _____
_____

2. What is the I.D. (inside diameter) of the gasket? _____
_____

3. How many holes does the gasket have? _____
_____

4. What is the angle between holes "a" and "b"? _____
_____

5. What is angle at "B"? _____
_____

6. What is the angle at "c"? _____
_____

7. What is the angle between holes "c" and "d"? _____
_____

8. What is the material used? _____
_____

9. What was the original size of the 2.25 (57.1) dia. hole? When the drawing was released? _____
_____

10. What is the thickness of this part? _____
_____

11. What is the width of the *groove?* _____

12. What is the B.C.? _____
_____

ONE VIEW DRAWING

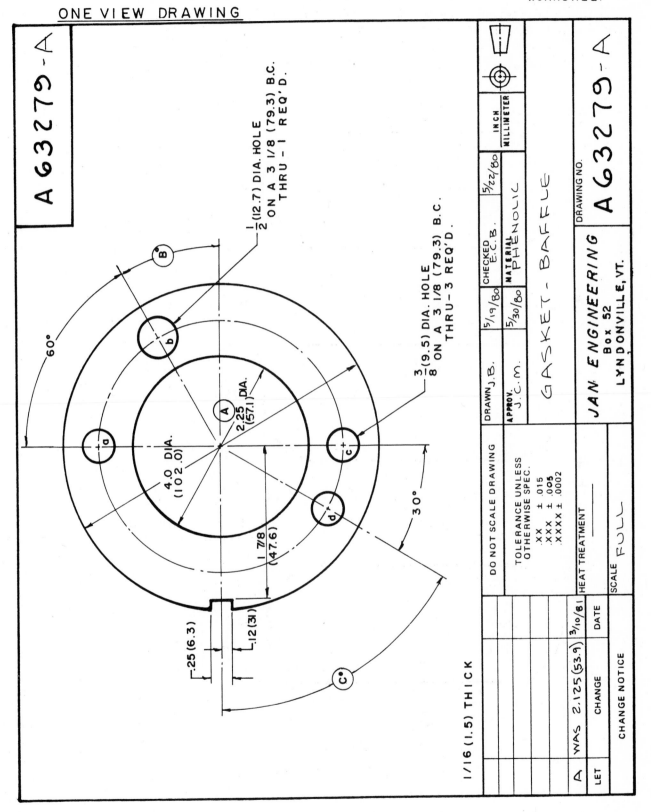

A 63279-A

1/2 (12.7) DIA. HOLE
ON A 3 1/8 (79.3) B.C.
THRU-1 REQ'D.

3/8 (9.5) DIA. HOLE
ON A 3 1/8 (79.3) B.C.
THRU-3 REQ'D.

60°

30°

B°

b

a

A

2.25 DIA.
(57.1)

4.0 DIA.
(102.0)

c

d

C°

1 7/8
(47.6)

.25 (6.3)

.12 (31)

1/16 (1.5) THICK

DO NOT SCALE DRAWING	DRAWN J.B.	5/19/80	CHECKED E.C.B.	5/22/80	INCH
					MILLIMETER
TOLERANCE UNLESS OTHERWISE SPEC.	APPROV. J.C.M.	5/30/80	MATERIAL PHENOLIC		
.XX ± .015 .XXX ± .005 .XXXX ± .0002					
	HEAT TREATMENT		GASKET-BAFFLE		
	SCALE FULL		JAN ENGINEERING		
			Box 52		
			LYNDONVILLE, VT.		
A WAS 2.125 (53.9) 3/10/81			DRAWING NO. A 63279-A		
LET CHANGE	DATE				
CHANGE NOTICE					

DIMENSIONING ROUNDED SHAPES

Rounded shapes are usually dimensioned from a center line. They must locate the radius swing point, and indicate the required radius. Below are three examples of ways a rounded shape may be dimensioned.

(A) Location of both swing points and radius.

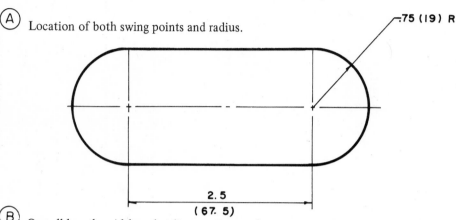

(B) Overall length, width and radius are given — (not swing points).

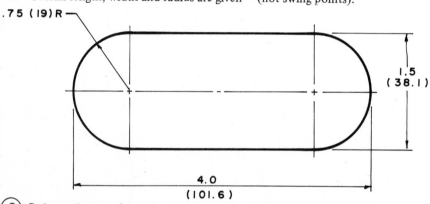

(C) Swing points are located, overall length and width but the radius is indicated by simply a "R"
— The radius is actually found in the center point location, .75 (19.0) dimension.

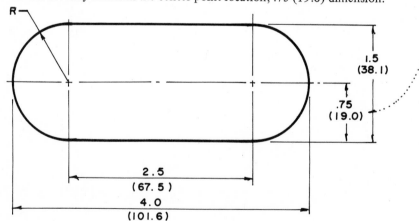

DIMENSIONING ROUNDED SHAPES WORKSHEET

Using the "given" example drawing *above* answer *'A'* through *'D'* in the spaces provided. Give answers in both systems. See end of chapter for answers.

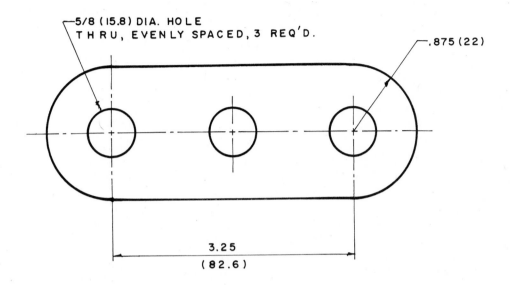

Questions:

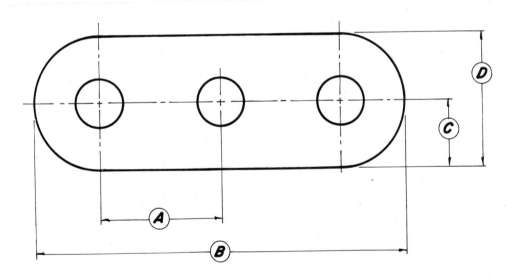

Answers:

TWO-VIEW DRAWINGS

Up to this point all drawings have been one-view illustrations of the object. In order to fully describe the objects in this section, more than one view must be used. Carefully study this pictorial illustration of a two-view drawing.

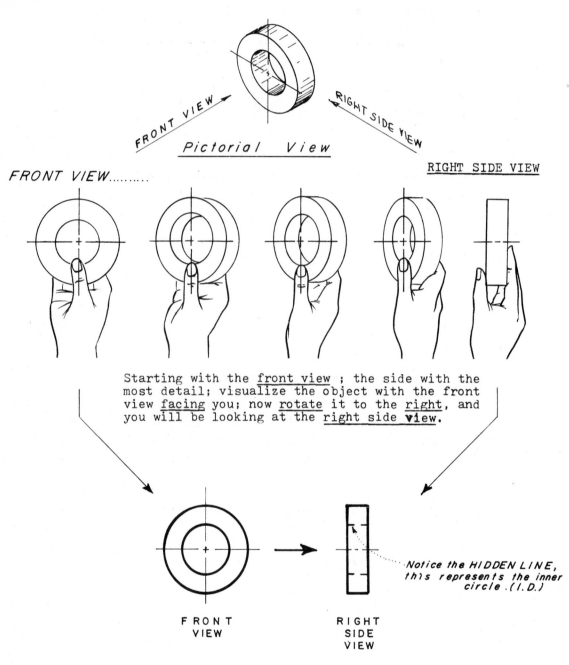

Pictorial View

FRONT VIEW..........

RIGHT SIDE VIEW

Starting with the front view ; the side with the most detail; visualize the object with the front view facing you; now rotate it to the right, and you will be looking at the right side view.

F R O N T
VIEW

R I G H T
SIDE
VIEW

*Notice the HIDDEN LINE,
this represents the inner
circle .(I.D.)*

Two-view representation

Notice, the *right side view* is directly to the *right* of the *front view.*

PROJECTION LINES

Projection lines are lines used by the draftsperson to construct the drawing. These lines are very light solid lines that do not copy on the blueprint. Projection lines are useful in transfering a surface from one view to another view in order to fully understand its shape, or location. Example: Dia. *"A"* in the *front view* is "projected" into the *right side view* and is represented by the hidden lines.

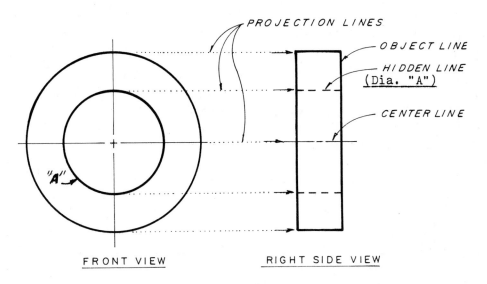

Note diameters *A, B, C,* and *D* in the *front view*, notice how they are represented in the *right side view*. Notice arc *'D'* (front view) is the hidden surface in the *right side view*. Project each arc from the *front view* to the *right side view*.

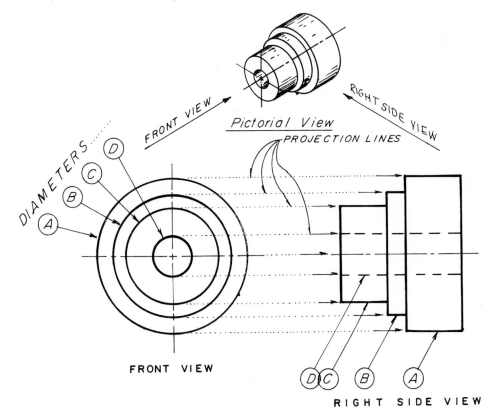

TWO-VIEW DRAWINGS WORKSHEET

Using drawing No. A332941 answer the following questions in the spaces provided. (Use both inch and millimeter systems). Add light 'projection' lines if it will help you answer any question. See end of chapter for answers.

1. What is the name of the part? _____

_____

2. What material is used in the manufacturing of this part? _____

_____

3. What kind of a line is: _A_ _____ _E_ _____

   _F_ _____ _G_ _____ _K_ _____

4. The diameter indicated by '_A_', is what dia.? _____

5. The diameter indicated by '_B_', is what dia.? _____

6. The diameter indicated by '_C_', is what dia.? _____

7. The diameter indicated by '_D_', is what dia.? _____

8. The diameter indicated by '_E_', is what dia.? _____

9. What is dimension '_H_'? _____

10. The diameter indicated by '_I_', is what dia.? _____

11. What is dimension '_J_'? _____

12. Calculate dimension '_L_'. _____

_____

13. Calculate the millimeter equivalent of dimension '_M_'. _____

_____

14. Convert this millimeter dimension '_N_' to inches. _____

_____

15. There is an important dimension missing, what is it? _____

_____

TWO-VIEW DRAWING

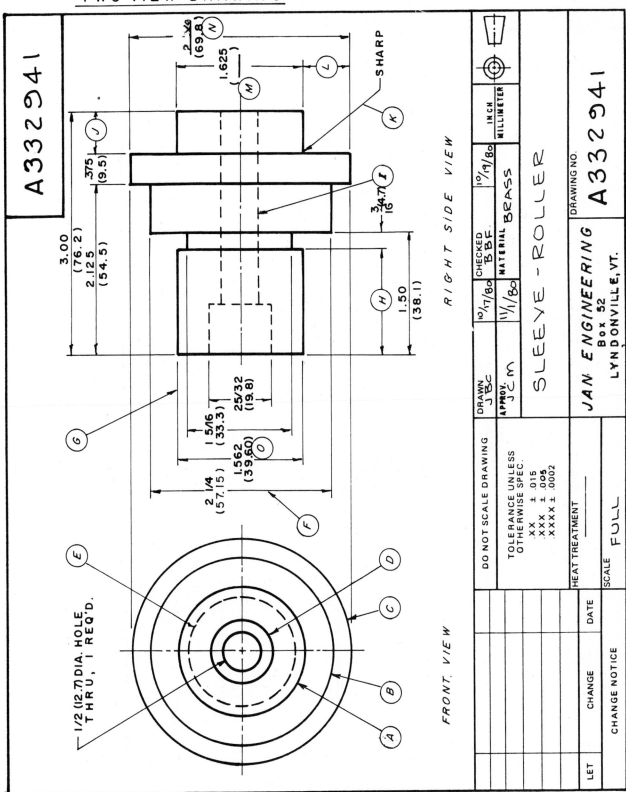

A332941

RIGHT SIDE VIEW

FRONT VIEW

SHARP

2 ¾ (69.8)

1.625

3.00 (76.2)

2.125 (54.5)

.375 (9.5)

3/16 (4.7)

1.50 (38.1)

1 5/16 (33.3)

25/32 (19.8)

1.562 (39.60)

2 1/4 (57.15)

1/2 (12.7) DIA. HOLE THRU, 1 REQ'D.

	INCH
	MILLIMETER

DRAWN	JBC	10/17/80	CHECKED	BBF.	10/19/80
APPROV.	JCM	11/1/80	MATERIAL	BRASS	

SLEEVE-ROLLER

JAN ENGINEERING
Box 52
LYNDONVILLE, VT.

DRAWING NO.

A332941

DO NOT SCALE DRAWING

TOLERANCE UNLESS
OTHERWISE SPEC.
.XX ± .015
.XXX ± .005
.XXXX ± .0002

HEAT TREATMENT _____

SCALE FULL

LET	CHANGE	DATE

CHANGE NOTICE

PROJECTION LINES

Projection lines are useful in determining what various lines represent. The illustrations so far have been with round shaped objects, projection lines are used with flat objects also.

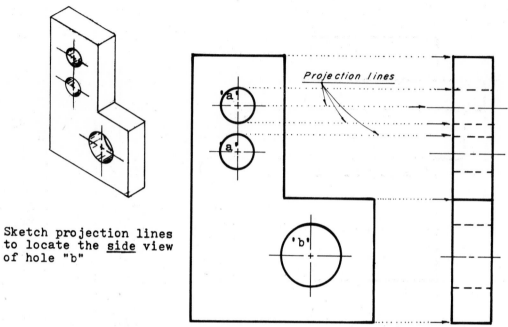

Sketch projection lines
to locate the <u>side</u> view
of hole "b"

FRONT VIEW   RIGHT SIDE VIEW

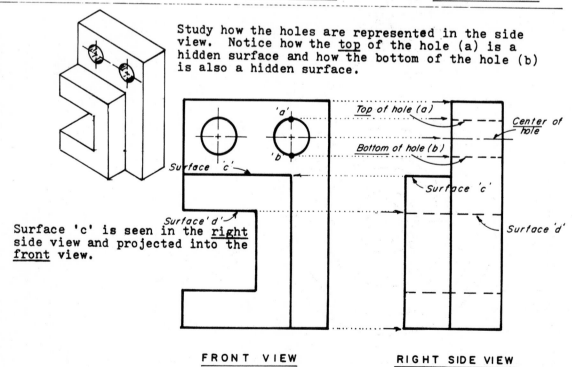

Study how the holes are represented in the side
view. Notice how the <u>top</u> of the hole (a) is a
hidden surface and how the bottom of the hole (b)
is also a hidden surface.

Surface 'c' is seen in the <u>right</u>
side view and projected into the
<u>front</u> view.

FRONT VIEW   RIGHT SIDE VIEW

TWO-VIEW DRAWINGS

A two-view drawing could be made up of the *front* view (most important) and a *right side* view or of the front view and a *top* view. This example will illustrate the latter.

Starting with the front view, the view with the most detail, visualize the object with the front view facing you, now *revolve* it toward the top, and you will be looking at the *top* view. Notice the top view is directly above the front view.

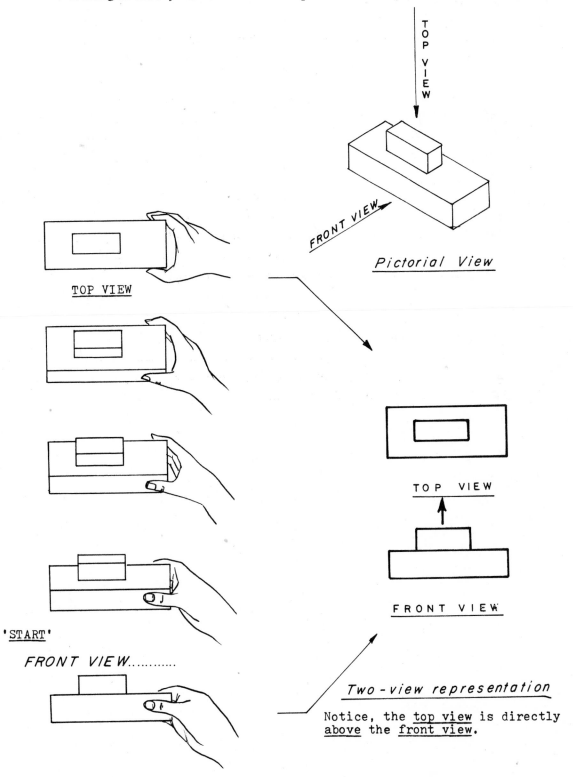

T O P V I E W

FRONT VIEW

Pictorial View

TOP VIEW

TOP   VIEW

FRONT   VIEW

'START'

FRONT VIEW............

Two-view representation

Notice, the <u>top view</u> is directly <u>above</u> the <u>front view</u>.

TWO-VIEW DRAWING WORKSHEET

Using plan A863578-A, answer the following questions. See end of chapter for answers.

1. Surface 'A' in the top view is what surface in the front view? _____

2. What is dimension 'B'? _____

3. Surface 'C' in the top view is what surface in the front view? _____

4. What does the hidden line at 'S' indicate? _____

5. What was the change made 10/9/80? _____

6. Calculate dimension 'F'. _____

7. Surface 'K' in the front view is what surface in the top view? _____

8. Surface 'N' in the front view is what surface in the top view? _____

9. Surface 'Q' in the top view is what surface in the front view? _____

10. What is the *width* in *millimeters* of the slot? _____

11. Surface 'R' in the front view represents what? _____

12. Surface 'T' in the front view is what surface in the top view? _____

TWO-VIEW DRAWINGS

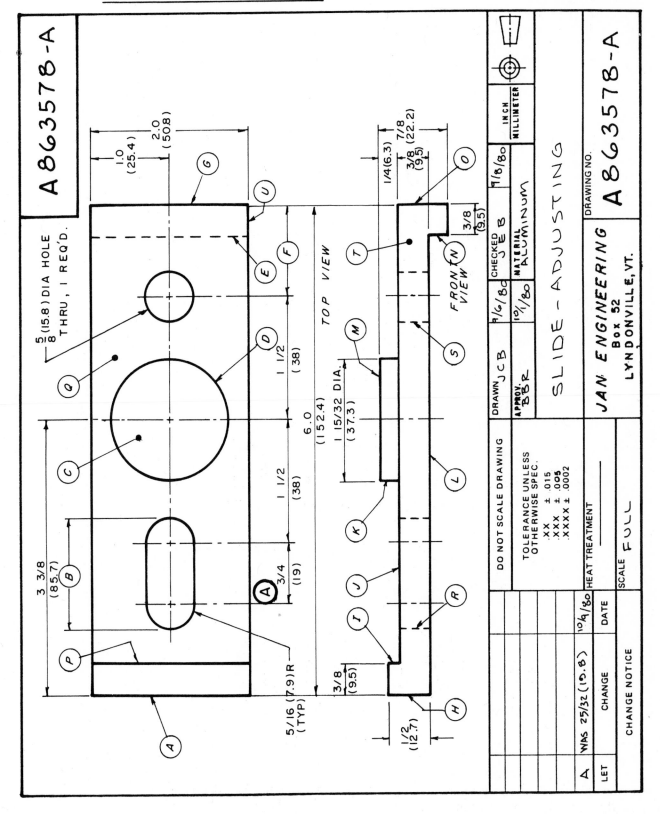

A863578-A

5/8 (15.8) DIA HOLE
THRU, 1 REQ'D.

1.0 (25.4)
2.0 (50.8)

3 3/8 (85.7)

6.0 (152.4)

1 1/2 (38)

1 1/2 (38)

3/4 (19)

A

5/16 (7.9) R (TYP)

TOP VIEW

1 15/32 DIA. (37.3)

FRONT VIEW

7/8 (22.2)
1/4 (6.3)
3/8 (9.5)
3/8 (9.5)
3/8 (9.5)
1/2 (12.7)

DO NOT SCALE DRAWING

TOLERANCE UNLESS OTHERWISE SPEC.
.XX ± .015
.XXX ± .005
.XXXX ± .0002

HEAT TREATMENT

SCALE FULL

DRAWN JCB
CHECKED JEB 7/6/80
APPROV. BBR 10/1/80
MATERIAL ALUMINUM
7/8/80
INCH / MILLIMETER

SLIDE-ADJUSTING

JAN ENGINEERING
Box 52
LYNDONVILLE, VT.

DRAWING NO.
A863578-A

A WAS 25/32 (19.8) 10/9/80
LET CHANGE DATE
CHANGE NOTICE

THREE-VIEW DRAWING

A three-view drawing is a combination of three or more views, usually a front view, top view, and a right side view. The object is revolved upwards toward you for the top view and rotated to the right for the right side view.

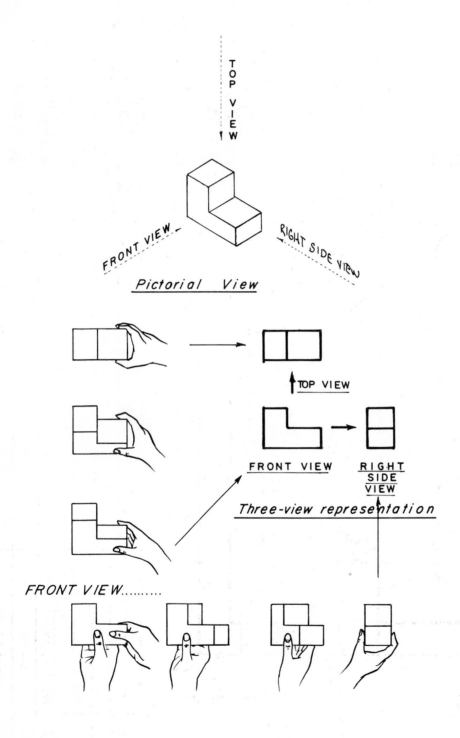

Notice: The *top view* is directly *above* the front view and the right side view is directly to the *right of* the front view.

MULTIVIEW SYSTEM

A multiview drawing is simply a drawing of an object, viewed from more than one place, i.e., looking directly at the front, *from* the front; looking directly at the right side, *from* the right side and so on. Each view of an object is in relationship to the next view but illustrates a different viewpoint.

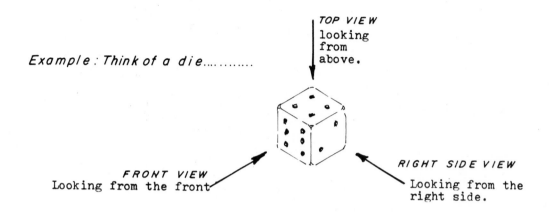

Example: Think of a die............

TOP VIEW
looking
from
above.

FRONT VIEW
Looking from the front

RIGHT SIDE VIEW
Looking from the
right side.

If you use the side of the die with six dots at the *front* view, (the side with the most detail) and look *only* at the *front*, this is what you will see:

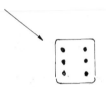

Look straight *down* from above at the die, this is what you will see:

Walk around to the *right* side of the die and it will look like this:

MULTIVIEW SYSTEM

Visualize what the die looks like:

Pictorial view

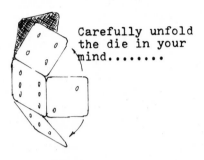

Carefully unfold
the die in your
mind........

U N F O L D

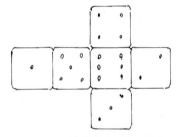

Now think of it flattened out..............

F L A T T E N

And then, separate the
flattened surfaces......

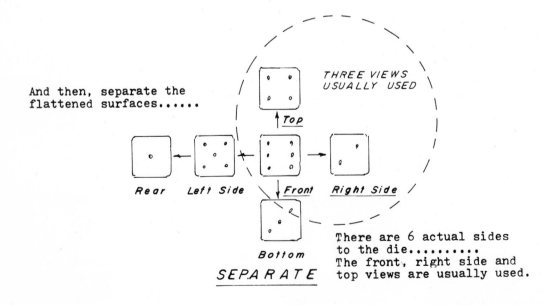

*THREE VIEWS
USUALLY USED*

Top

Rear   *Left Side*   *Front*   *Right Side*

Bottom

S E P A R A T E

There are 6 actual sides
to the die.........
The front, right side and
top views are usually used.

MULTIVIEW DRAWINGS REVIEW

Referring back to the die:

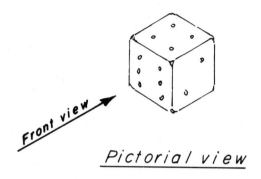

Pictorial view

Looking directly at the front of it you will see:

Front view

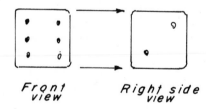

Front view   *Right side view*

Directly to the *right* of the front view, is the *right* side view.

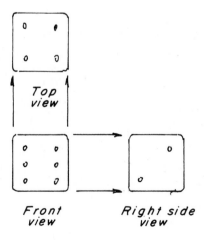

Top view

Front view   *Right side view*

Directly *above* the front view is the *top* view this constitutes the usual three views used in industry.

Width, Height, and Depth

To clarify any question with width, height, and depth, all references will be as illustrated:

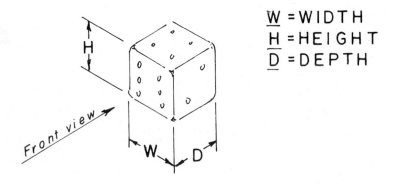

PROJECTION LINES

Projection lines are used in three-view drawings also. From the front view, lines project to the right to the right side view. From the front view lines project up to the top view. Don't forget, these projection lines are actually not seen on the white print copy, so if something is not clear you should lightly add them.

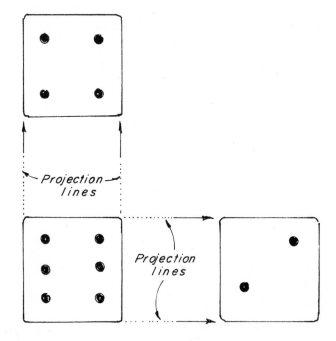

In three-view drawings, projection lines also go from the right side view *to* the top view, or vica versa. Study this illustration. Notice how surface 'A' in the top view is also surface 'A' in the right side view (study pictorial view if this is not clear). To draw these projection lines, draw a line *up* from surface 'A' in the right side view. Draw a line to the right, in line with surface 'A', from the top view to locate point "X." From this point ("X") draw a 45° angle (approx.) and project all lines as illustrated.

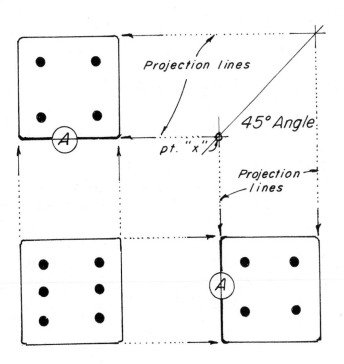

THREE-VIEW DRAWING WORKSHEET

Using the three-view drawing on p. 77 make sure you note and fully understand each point.

1. Surface 'A' in the front view is projected *up* to surface 'A' in the top view. _____

2. Surface 'B' in the top view is projected *down* to surface 'B' in the front view. _____

3. Locate surface 'C' in the right side view, project it *up* to the 45° angle and over to the top view—locate surface 'C' in the top view. _____

4. Locate surface 'C' in the top view and project it *down* to the front view. _____

5. Notice surface 'D' in the front view and where it is located in the right side view. _____

6. Surface 'E' in the right side view is projected *up/over* to the top view. _____

7. Notice surface 'F' in the front, top and right side views. _____

8. Project surface 'G' from the front view up to the top view and over to the right side view. Note the width of 'G' is exactly the same width in the top view. _____

9. Carefully study surface 'H'. Note that it is projected from the *top* view over/down to the right side view—then over to the front view. _____

10. Locate surface 'I' in the front view, now locate it in the top view from the right side view. _____

11. Notice, surface 'J' in the front view, it appears as a hidden surface in the right side view. _____

Before beginning, study this three-view drawing. Note the position of the *front view,* the right side view and the top view. Carefully notice all projection lines, especially projecting from the right side view, up to and through the 45° angle and over to the top view. Try to get a mental picture of what this object would look like if you could hold it in your hands.

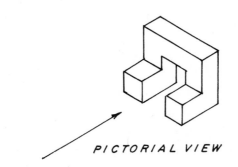

PICTORIAL VIEW

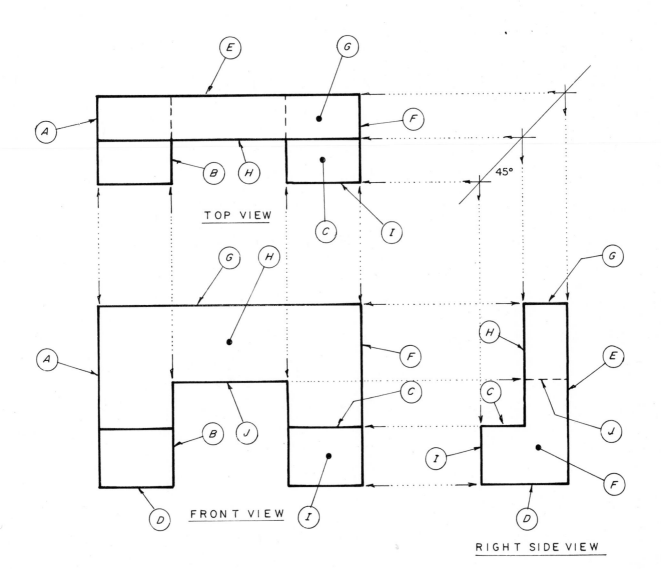

THREE-VIEW DRAWINGS WORKSHEET I

Using the three-view drawing below, answer the following questions in the spaces provided. See end of chapter for answers.

1. What is dimension 'A'? _____

    _____

2. Calculate dimension 'B'? _____

    _____

3. How many millimeters is dimension 'C'? _____

    _____

4. Assuming the 7/8 (22.2) notch is in the center of the part, what would dimension 'D' be? _____

    _____

5. Surface 'E' in the top view is what surface in the right side view? _____

    _____

6. Surface 'F' in the right side view is what surface in the top view? _____

    _____

7. Surface 'G' in the right side view is what surface in the top view? _____

    _____

8. Surface 'H' in the front view is what surface in the top view? _____

    _____

9. Surface 'I' in the right side view is what surface in the top view? _____

    _____

10. What kind of a line is 'Q'? _____

    _____

11. What scale is used to draw the part? _____

    _____

12. What is the length, height, and depth of the part? _____

    _____

THREE VIEW DRAWINGS

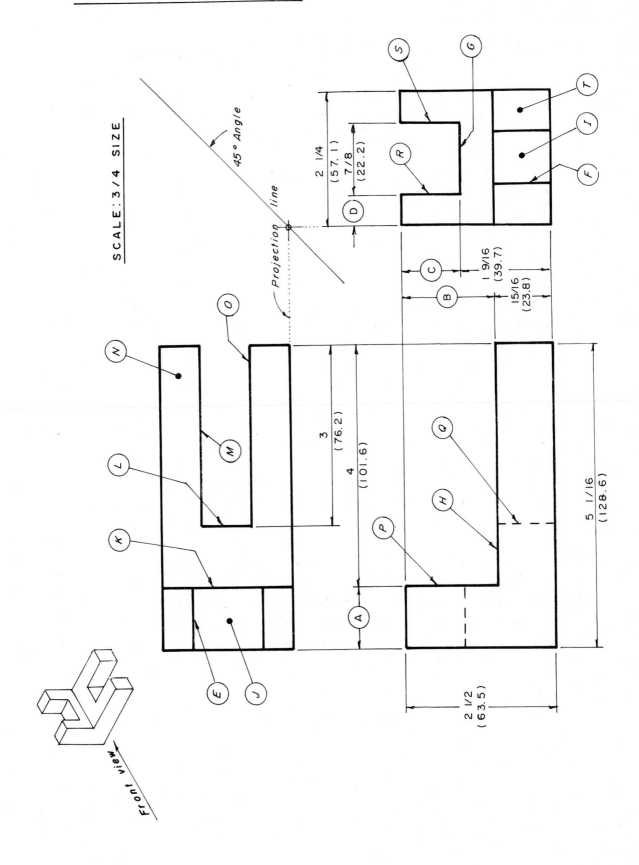

SCALE: 3/4 SIZE

45° Angle

Projection line

Front view

THREE-VIEW DRAWINGS WORKSHEET II

Using the three-view drawing below, answer the following questions in the spaces provided. See end of chapter for answers.

1. Calculate dimension 'A'. _____

_____

2. What is dimension 'B'? _____

_____

3. What is the *length, height,* and *depth* of this object? _____

_____

4. Dimension 'C' is equal to? _____

_____

5. What is dimension 'D'? _____
6. Calculate dimension 'E'. _____
7. What is dimension 'F'? _____
8. What is dimension 'G'? _____
9. Dimension 'H' is equal to? _____
10. Line 'L' in the top view is what surface in the front view? _____

_____

11. Line 'Q' in the front view is what surface in the top view? _____

_____

12. Line 'T' in the front view is what surface in the right side view? _____

_____

13. Hidden line 'Z' in the right side view is what surface in the top view? _____

_____

14. Surface 'W' in the right side view is what surface in the front view? _____

_____

15. Surface 'W' in the right side view is what surface in the top view? _____

_____

16. Surface 'J' in the top view represents what surface in the right side view?

_____

17. Line 'U' in the rigth side view represents what surface in the top view? _____

_____

18. In the top view, a radii is noted by a 'R'. What is this radius? _____

_____

THREE-VIEW DRAWINGS

WORKSHEET

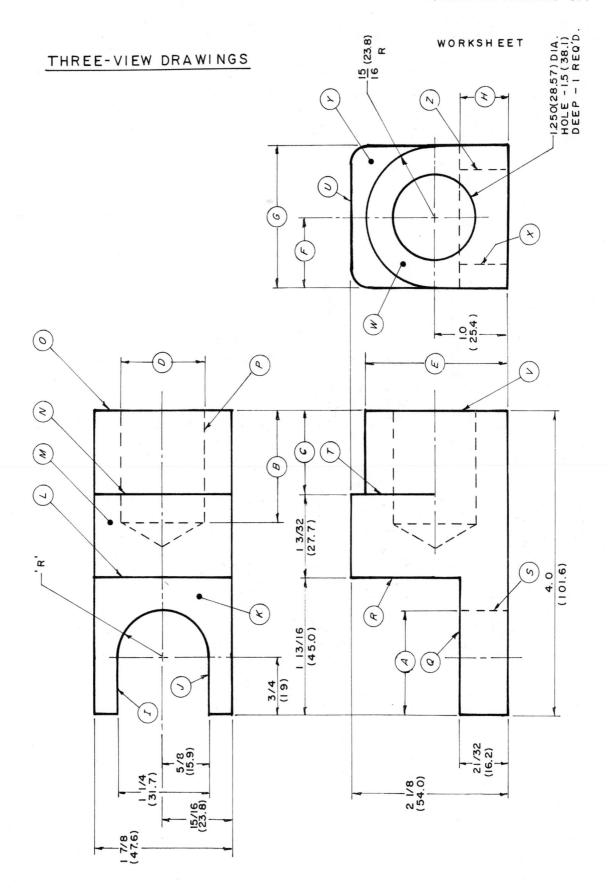

South St. Paul Public Library
106 Third Avenue North
South St. Paul, MN 55075

CALL-OFFS AND REPRESENTATIONS

Carefully study each standard type of hole used in industry, note its representation and its "call-off." The *process* used to make a *hole* should not be noted in the "call-off" on the drawing. All kinds of holes must be *drilled first* regardless of the *final* process, i.e., a reamed hole is drilled before reaming, etc. The difference between a drilled hole and a reamed hole is the *tolerance*.

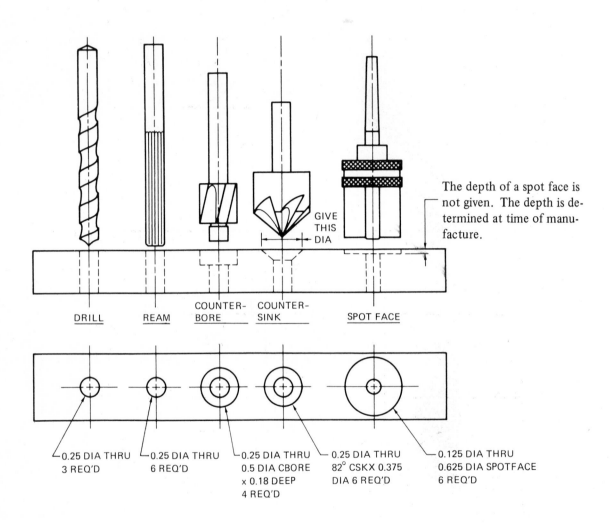

GIVE THIS DIA

The depth of a spot face is not given. The depth is determined at time of manufacture.

DRILL   REAM   COUNTER-BORE   COUNTER-SINK   SPOT FACE

0.25 DIA THRU
3 REQ'D

0.25 DIA THRU
6 REQ'D

0.25 DIA THRU
0.5 DIA CBORE
x 0.18 DEEP
4 REQ'D

0.25 DIA THRU
82° CSK X 0.375
DIA 6 REQ'D

0.125 DIA THRU
0.625 DIA SPOTFACE
6 REQ'D

THREE-VIEW DRAWINGS WORKSHEET III

Using drawing A395473, answer the following questions. See end of chapter for answers.

1. What is dimension 'A'? _____

_____

2. What is dimension 'B' and *what* is it called? _____

_____

3. Calculate dimension 'D'? _____

_____

4. What is the size of 'F'? _____

_____

5. Calculate dimension 'I.' _____

_____

6. Calculate dimension 'J.' _____

_____

7. Surface 'E' in the top view is what surface in the right side view? _____

_____

8. Surface 'K' in the top view is what surface in the right side view? _____

_____

9. What is the height of the object at 'L'? _____

_____

10. Surface 'M' in the front view is what surface in the top view? _____

_____

11. Center line 'O' is the center of what? _____

_____

_____

12. Surface 'P' is the front view is what surface in the right side view? _____

_____

_____

13.   What is the size of dimension 'Q'? _____
_____
_____
_____

14.   Surface 'R' in the front view is what surface in the top view? _____
_____
_____
_____

15.   Surface 'S' in the front view is what surface in the top view? _____
_____
_____
_____

16.   What is the dimension of 'T'? _____
_____
_____
_____

17.   What is the dimension of 'U'? _____
_____
_____
_____

18.   The wavy line under a dimension (w) indicates what? _____
_____
_____
_____

19.   Surface 'V' in the right side view is what surface in the front view? _____
_____
_____
_____

20.   What is the width of the object? _____
_____
_____
_____

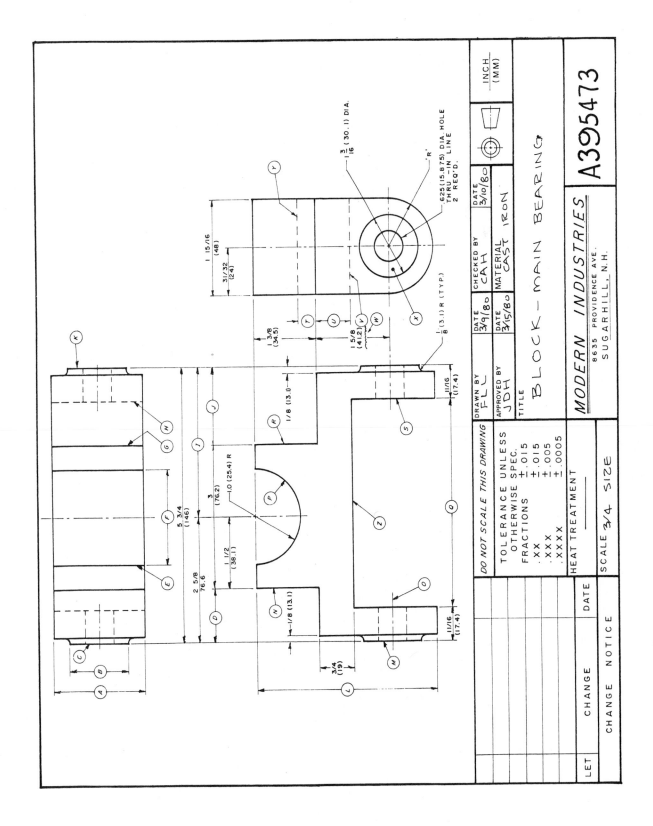

EVALUATION

Using drawing A374593-A answer the following questions. See end of chapter for answers.

1. What system of dimensioning is used on this drawing? _____
   _____

2. What is the width, height, and depth of this object? _____
   _____

3. What is dimension 'A'? _____

4. Surface 'C' in the top view is what surface in the right side view? _____
   _____

5. Calculate radius 'D'? _____

6. Surface 'G' in the top view is what surface in the right side view? _____
   _____

7. Surface 'F' in the top view is what surface in the right side view? _____
   _____

8. Calculate dimension 'H'? _____

9. Surface 'I' in the front view is what surface in the right side view? _____
   _____

10. Surface 'K' in the front view is what surface in the top view? _____
    _____

11. Hole 'a' is represented in the right side view by what letter? _____
    _____

12. What is dimension 'M'? _____

13. What was dimension 'O' before the change? _____

14. Calculate dimension 'P'. _____

15. 'R' is the center line of what? _____

16. What does the wavy line at 'S' indicate? _____

17. What is the radii at 'Z'? _____

18. The hidden line at 'N' represents what? _____

19. What are the upper/lower limits for the center to center of the holes distance?
    _____

20. What is the largest size the 5/8 (15.8) dia. hole can be made and pass inspection? _____

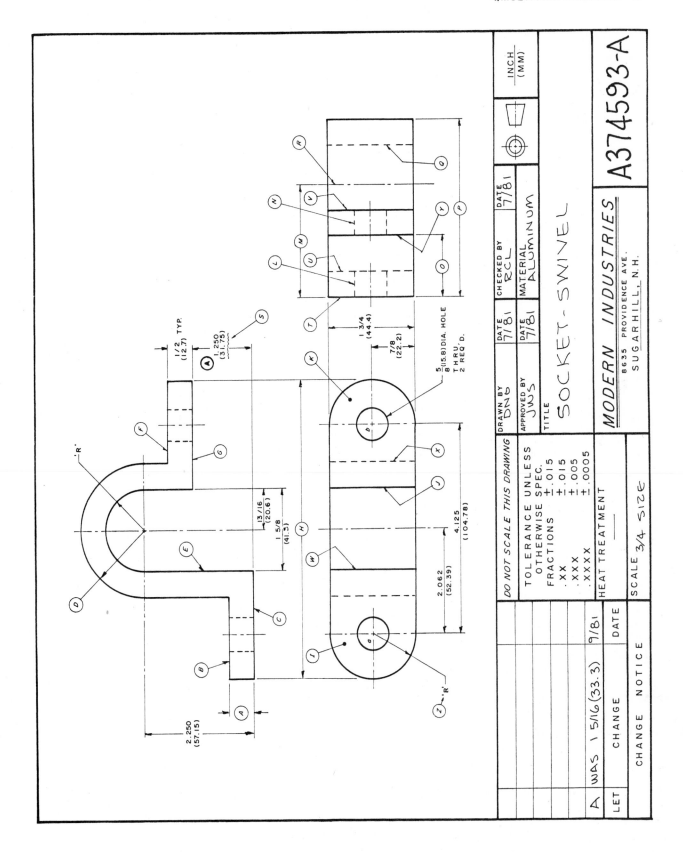

ANSWERS TO TERMS TO KNOW AND UNDERSTAND

Aligned—All dimensions are aligned with the dimension line and read from the *bottom* and *right* side of the drawing.

Unidirectional—All dimensions are read from the *bottom* of the drawing (uni- means one).

Out-Of-Scale Dimensions—Have a wavy line under them to note they are *not to* scale.

Metric Diametral Dimension Symbol—φ

One View Drawing—A drawing with one view required to represent the object.

Bolt Circle—A dia. holes are located on.

Two View Drawings—A drawing with two views required to represent the object.

Projection Lines—Used to determine what line or surface in one view, is the same line or surface in another view. Lines *not* on the drawing.

Three View Drawings—A drawing with three views required to represent the object.

Width, Height, And Depth—Three dimensions at visual perception.

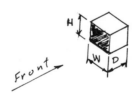

Kinds Of Holes—Drilled, reamed, counter-bored, counter-sunk, and spotfaced.

Boss Or Pad—A raised section, usually round but could be square or rectangular.

ONE VIEW DRAWING WORKSHEET I ANSWERS

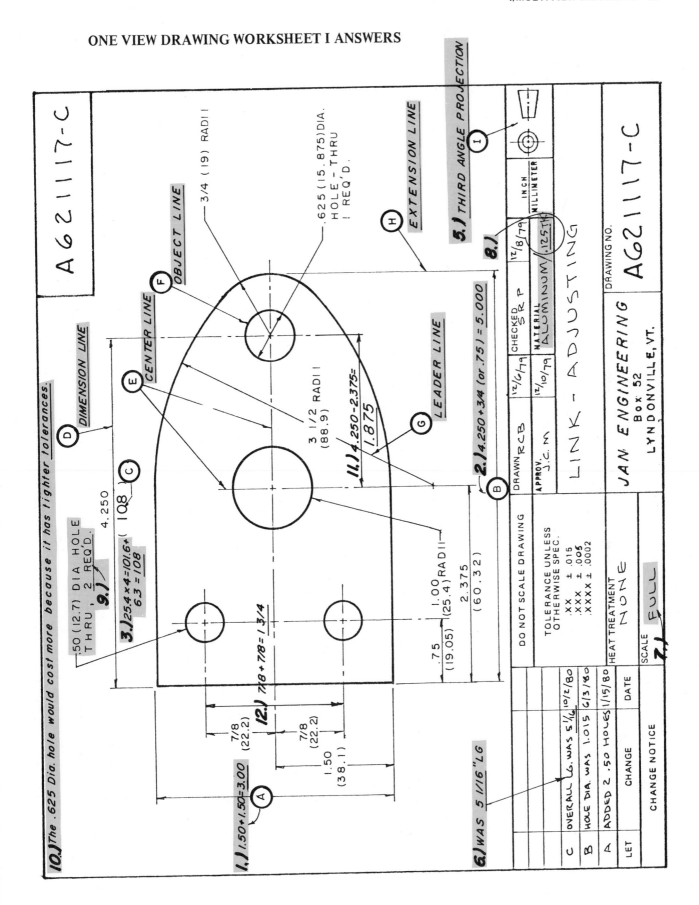

ONE VIEW DRAWING WORKSHEET II ANSWERS

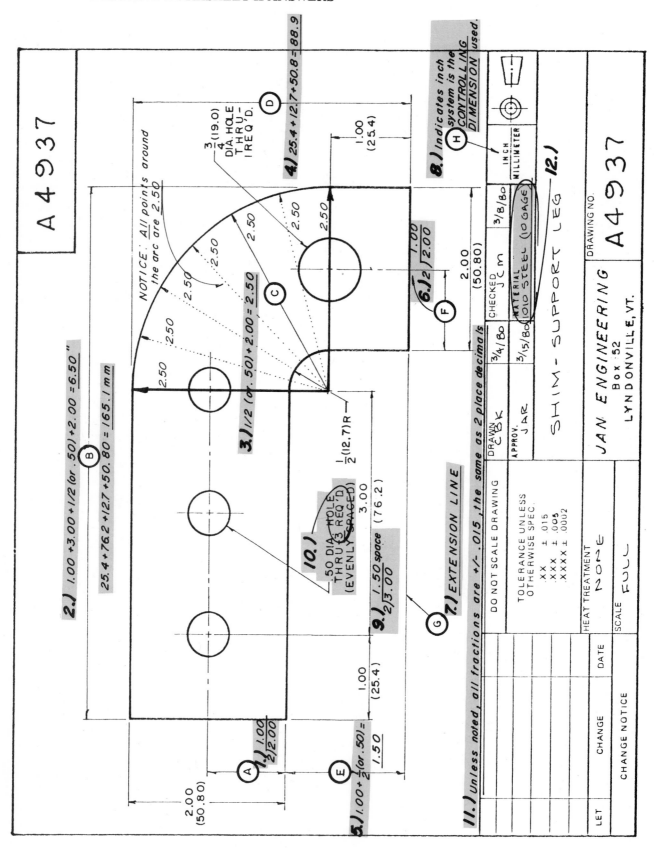

A4937

2.) 1.00 + 3.00 + 1/2 (or .50) + 2.00 = 6.50 "

25.4 + 76.2 + 12.7 + 50.80 = 165.1 mm

B

NOTICE: All points around the arc are 2.50

4.) 25.4 + 12.7 + 50.8 = 88.9

D

3/4 (19.0) DIA. HOLE THRU– 1 REQ'D.

2.50
2.50
2.50
2.50
2.50
2.50
2.50
2.50
2.50

C

3.) 1/2 (or .50) + 2.00 = 2.50

1/2 (12.7) R

1.00 (25.4)

6.12) 1.00 / 2.00

F

2.00 (50.80)

8.) Indicates inch system is the CONTROLLING DIMENSION used.

H

INCH MILLIMETER

3/8/80

CHECKED   J C m

MATERIAL 1010 STEEL (10 GAGE)   **12.)**

.50 DIA. HOLE THRU 3 REQ'D (EVENLY SPACED)   **10.)**

3.00 (76.2)

9.) 1.50 space / 2) 3.00

7.) EXTENSION LINE

G

1.00 (25.4)

1.) 1.00 / 2) 2.00

A

5.) 1.00 + 1/2 (or .50) = 1.50

E

2.00 (50.80)

11.) Unless noted, all fractions are +/- .015 , the same as 2 place decimals

DO NOT SCALE DRAWING

DRAWN   C B K   3/4/80

APPROV.   JAR   3/15/80

SHIM - SUPPORT LEG

JAN ENGINEERING   Box 52   LYNDONVILLE, VT.

DRAWING NO.   A4937

TOLERANCE UNLESS OTHERWISE SPEC.
.XX ± .015
.XXX ± .005
.XXXX ± .0002

HEAT TREATMENT   NONE

SCALE   FULL

LET   CHANGE   DATE

CHANGE NOTICE

ONE VIEW DRAWING WORKSHEET III ANSWERS

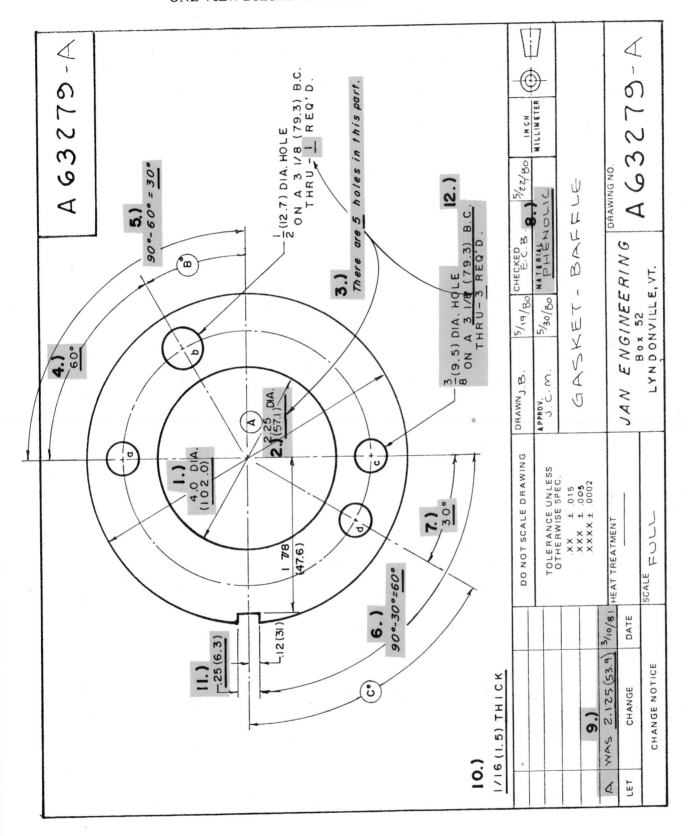

DIMENSIONING ROUNDED SHAPES WORKSHEET ANSWERS

Using the "given" example drawing *above;* Answer *'A'* through *'D' below* in the spaces provided. Give answers in both systems.

Given:

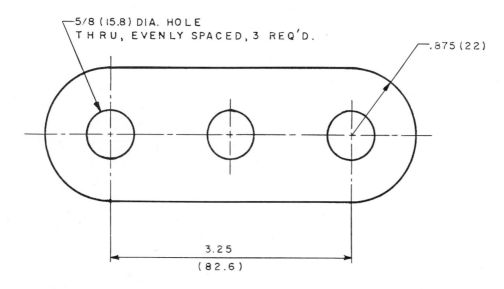

Questions:

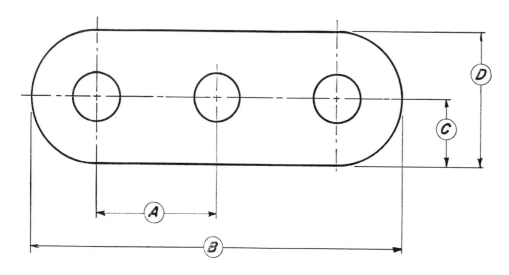

Answers:

$$\underline{A} \quad 2\overline{)3.25}^{\,1.625} \quad 1.625 \; (1\,5/8)$$
41.2

$$\underline{B} \quad \begin{array}{r} 3.250 \\ +\ .875 \\ +\ .875 \\ \hline 5.000 \end{array}$$
(127.0)

$$\underline{C} \quad (\textit{SAME AS RADIUS}) \quad .875$$
(22)

$$\underline{D} \quad \begin{array}{r} .875 \\ +\ .875 \\ \hline \end{array} \quad 1.75 \; (1\,3/4)$$
(44.4)

TWO-VIEW DRAWING WORKSHEET ANSWERS

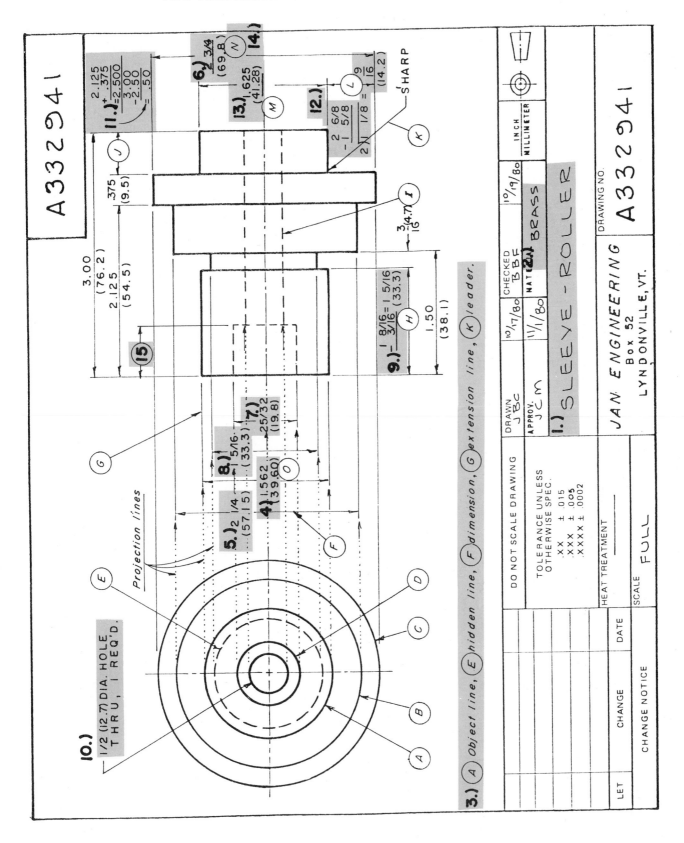

South St. Paul Public Library
106 Third Avenue North
South St. Paul, MN 55075

TWO-VIEW DRAWING WORKSHEET ANSWERS

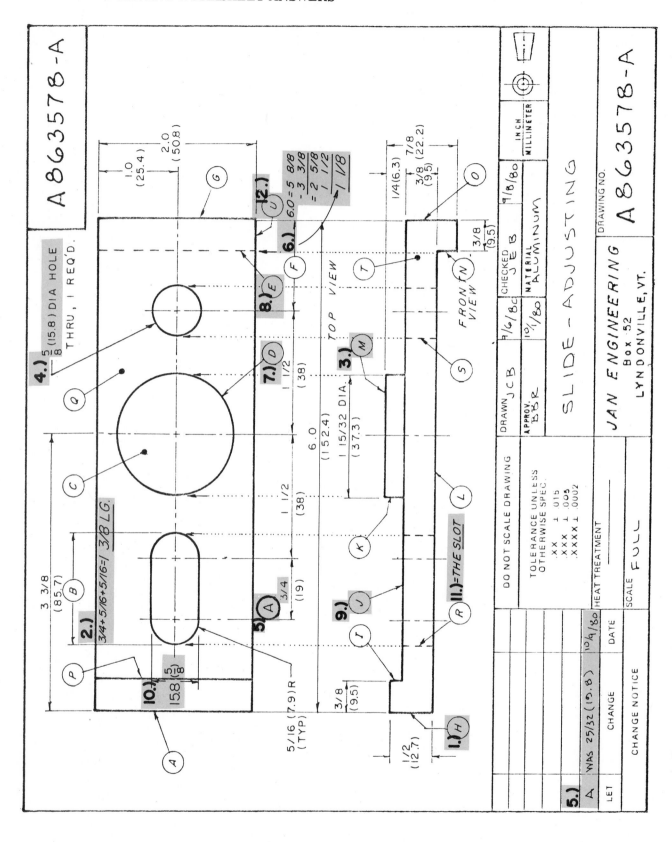

THREE-VIEW DRAWINGS WORKSHEET I ANSWERS

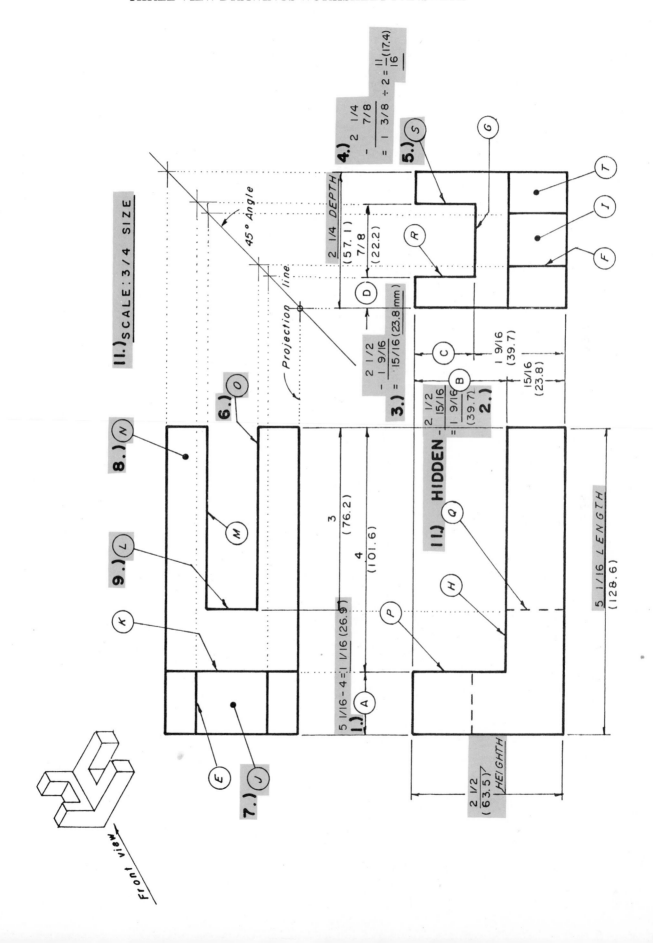

THREE-VIEW DRAWINGS WORKSHEET III ANSWERS

(See page 99 for a pictorial view, if you have
any problems visualizing this drawing).

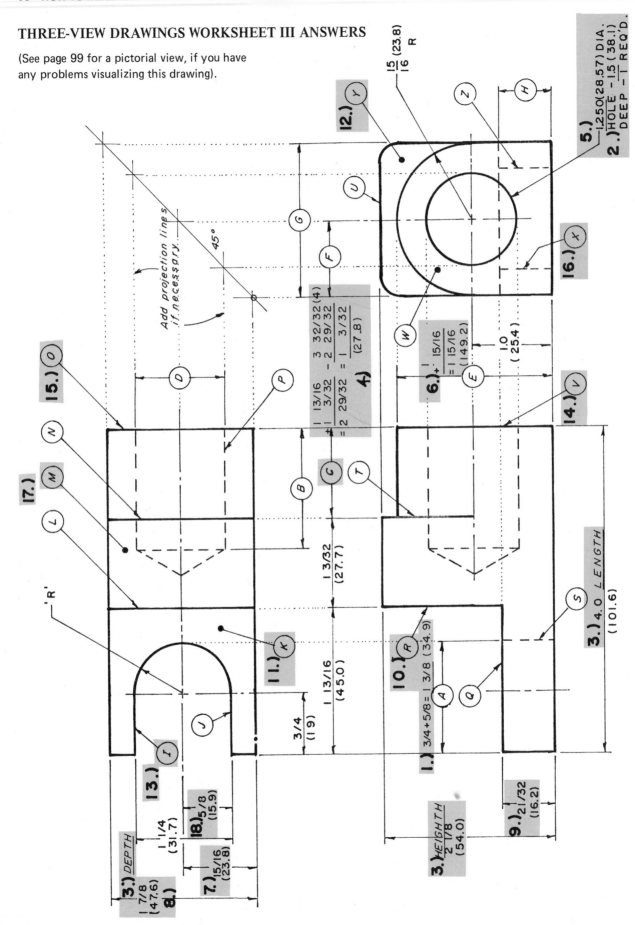

THREE-VIEW DRAWINGS WORKSHEET III ANSWERS

(See page 99 for a pictorial view, if you have any problems visualizing this drawing).

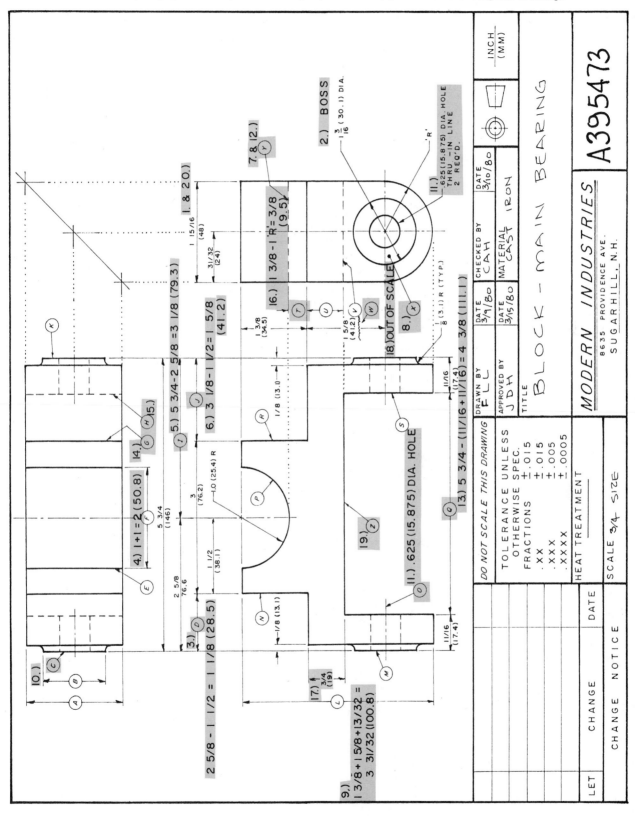

2.) BOSS

1.) .625 (15.875) DIA. HOLE THRU—IN LINE 2 REQ'D.

7. & 12.)

1. & 2 O.)

16.) 1 3/8 - 1 R = 3/8 (9.5)

8.) OUT OF SCALE

5.) 5 3/4 - 2 5/8 = 3 1/8 (79.3)

6.) 3 1/8 - 1 1/2 = 1 5/8 (41.2)

4.) 1 + 1 = 2 (50.8)

13.) 5 3/4 - (11/16 + 11/16) = 4 3/8 (111.)

11.) .625 (15.875) DIA. HOLE

15.)

14.)

3.)

17.) 3/4 (19.)

9.) 1 3/8 + 1 5/8 + 13/32 = 3 31/32 (100.8)

19.)

10.)

INCH (MM)

DRAWN BY FLL
APPROVED BY JDH
CHECKED BY CAH DATE 3/9/80
DATE 3/15/80 MATERIAL CAST IRON
DATE 3/10/80

TITLE BLOCK—MAIN BEARING

MODERN INDUSTRIES A395473
8635 PROVIDENCE AVE.
SUGARHILL, N.H.

DO NOT SCALE THIS DRAWING

TOLERANCE UNLESS OTHERWISE SPEC.
FRACTIONS ±.015
.XX   ±.015
.XXX ±.005
.XXXX ±.0005

HEAT TREATMENT

SCALE 3/4 SIZE

LET   CHANGE   DATE
CHANGE NOTICE

EVALUATION ANSWERS

(See page 99 for a pictorial view, if you have any problems visualizing this drawing).

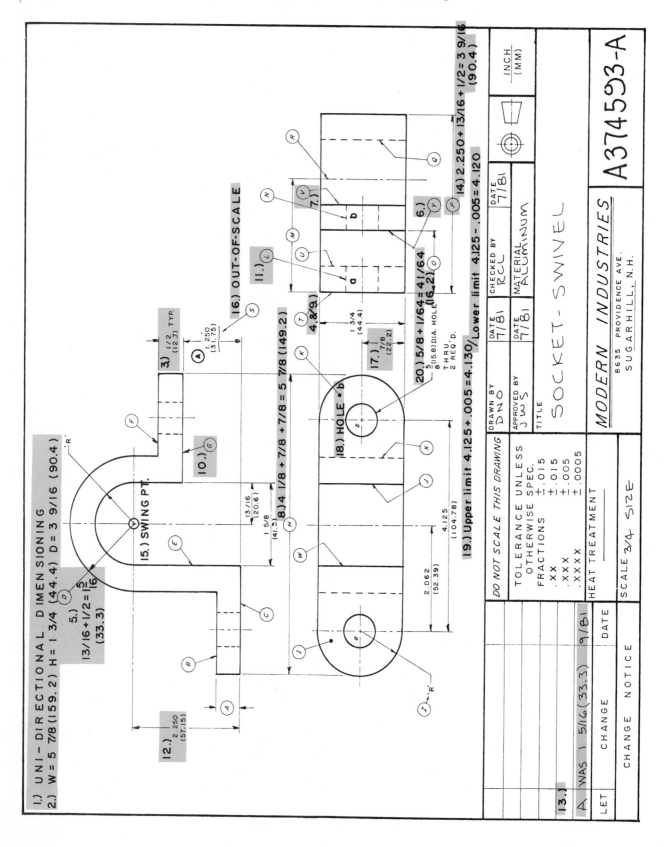

1.) UNI-DIRECTIONAL DIMENSIONING
2.) W = 5 7/8 (159.2) H = 1 3/4 (44.4) D = 3 9/16 (90.4)
3.) 1/2 (12.7) TYP.
4.8/9.)
5.) 13/16 + 1/2 = 1 5/16 (33.3)
6.)
7.)
10.)
11.) OUT-OF-SCALE
12.) 2.250 (57.15)
13.)
14.) 2.250 + 13/16 + 1/2 = 3 9/16 (90.4)
15.) SWING PT.
16.) OUT-OF-SCALE
17.) 7/8 (22.2)
18.) HOLE "b"
19.) Upper limit 4.125 + .005 = 4.130/Lower limit 4.125 – .005 = 4.120
20.) 5/8 + 1/64 = 41/64 (16.2)

5/8 (15.8) DIA. HOLE
THRU
2 REQ'D.

1 3/4 (44.4)

1.250 (31.75)

13/16 (20.6)

1 5/8 (41.5)

2.062 (52.39)

4.125 (104.78)

DO NOT SCALE THIS DRAWING

TOLERANCE UNLESS OTHERWISE SPEC.
FRACTIONS ± 1/64
.XX ± .015
.XXX ± .005
.XXXX ± .0005

HEAT TREATMENT

LET	CHANGE	DATE
A	WAS 1 5/16 (33.3)	9/81

CHANGE NOTICE

DRAWN BY DNO   DATE 7/81
CHECKED BY RCL   DATE 7/81
APPROVED BY JWS   DATE 7/81

MATERIAL ALUMINUM

TITLE SOCKET-SWIVEL

MODERN INDUSTRIES
8635 PROVIDENCE AVE.
SUGARHILL, N.H.

A374593-A

INCH / (MM)

SCALE 3/4 SIZE

THREE-VIEW DRAWINGS WORKSHEET II ANSWERS

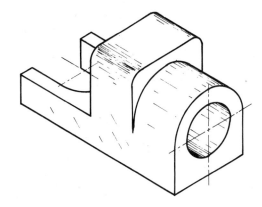

Pictorial view (from page 96)

THREE-VIEW DRAWINGS WORKSHEET III ANSWERS

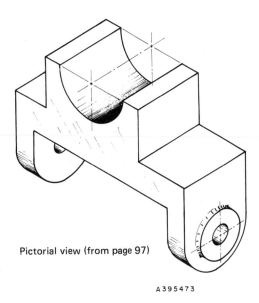

Pictorial view (from page 97)

A395473

THREE-VIEW DRAWINGS WORKSHEET III ANSWERS

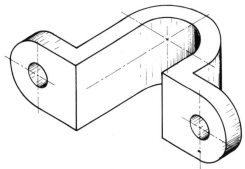

Pictorial view (from page 98)

A374593-A

UNIT 5

SECTION
VIEWS

Objective: To identify and practice interpreting the eight major kinds of section views and various drafting "practices."

TERMS TO KNOW AND UNDERSTAND

See end of chapter for answers.

Cutting Plane Line _____

Section Lining _____

Symbols Used To Represent Various Materials _____

Kinds Of Sections: _____

    Full Section _____

    Half Section _____

    Broken-out Section _____

    Offset Section _____

    Rotated Section _____

    Assembly Section _____

    Conventional Breaks _____

'Fillets'_____

Rounds_____

'Run-outs'_____

SECTION VIEWS

A drawing is usually represented by two or more views; usually a front view, top view, and/or right side view. All hidden surfaces are illustrated by hidden lines. Sometimes however, there are so many hidden lines a section view must be used to fully describe the object. There are various kinds of section views, each will be illustrated, but there *are a few points* common to all.

Cutting Plane Line

Is an imaginary cut *through* the object. Think of it as a saw cutting across the object. The cutting plane line appears as: ▬ ▬ ▬ ▬ ▬ ▬ ▬ ▬

or ▬▬ ▬ ▬ ▬▬▬ ▬ ▬ ▬▬

This is a *thick* line.

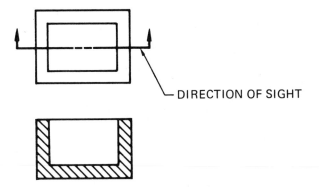

Direction Of Sight

There are arrows at the ends of a cutting plane line. These arrows point in the direction you are to view the section view.

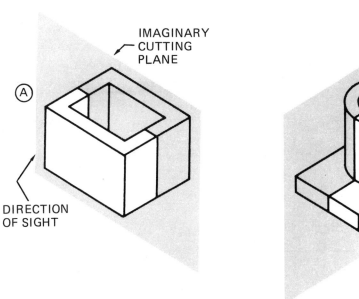
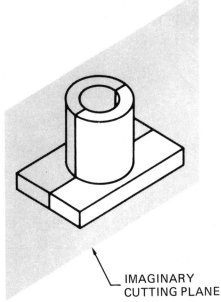

Section Lining

Is a *thin* line, usually at a 45° angle that denotes exactly where, and only where, the cutting plane line passes through. The new standard is illustrated to the *left* (all purpose) and is used for all materials. The old standard used various lines, as illustrated, to indicate the exact material used on the part.

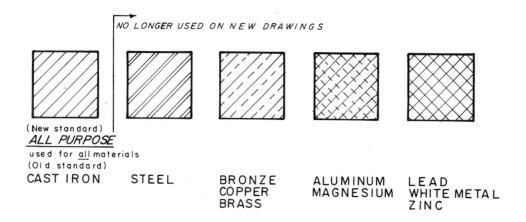

NO LONGER USED ON NEW DRAWINGS

(New standard)
ALL PURPOSE
used for all materials
(Old standard)

CAST IRON   STEEL   BRONZE   ALUMINUM   LEAD
                        COPPER   MAGNESIUM   WHITE METAL
                        BRASS   ZINC

Notice how the section lining is placed *only* where the cutting plane line passes through the object.

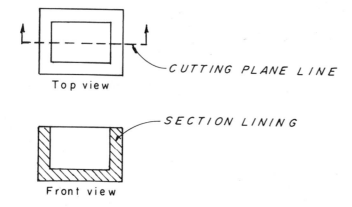

Top view

CUTTING PLANE LINE

SECTION LINING

Front view

KINDS OF SECTION VIEWS

The major kinds of sections are noted below—each will be explained.

- Full
- Half
- Broken-out
- Offset
- Rotated (revolved)
- Removed
- Assembly
- Thin-wall

FULL SECTION

A full section view is simply a view that is cut fully through. A full section view simply cuts through the object so the interior can be seen much easier. Study the illustrations below.

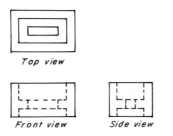

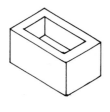

R E G U L A R   3 - V I E W   D R A W I N G
Notice hidden lines in front view.

P I C T O R I A L   V I E W
<u>Rule</u>: Hidden lines are usually omitted
from the view that appears
in section.

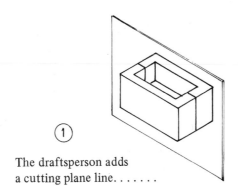

① The draftsperson adds
a cutting plane line. . . . . . .

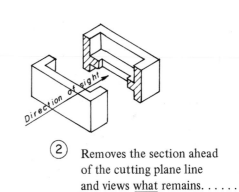

② Removes the section ahead
of the cutting plane line
and views <u>what</u> remains. . . . . .

③ Converting this back to a regular 3-view drawing, the <u>top</u> view contains
the cutting plane line, the front view now illustrates <u>what</u> will be seen
viewing the object in the directions indicated by the arrows.

Section lining is
placed <u>only</u> on the
surface the cutting
plane line passed through.

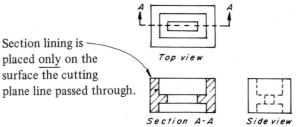

F R O N T   V I E W   A S   A   F U L L   S E C T I O N

FULL SECTION WORKSHEET

Using plan #A70031 answer the following questions in the space provided. (Note the cutting plane line A-A). See end of chapter for answers.

1. What scale is this drawing drawn to? _____

2. What is dimension 'A'? _____

3. Surface 'D' in the top view is what surface in the right side view? _____
   _____

4. Calculate dimension 'F'. _____

5. Calculate dimension 'G'! _____

6. Calculate dimension 'H'! _____

7. Surface 'I' in the front view is what surface in the top view? (Do not forget the front view is now a full section view). _____

8. Surface 'J' in the front view is what surface in the right side view? _____
   _____

9. Surface 'K' in the front view is what surface in the top view? _____
   _____

10. How deep is the counterbore? _____

11. What is the material used to make this part? _____

12. How many 1/2 (12.7) dia. holes are there? _____

FULL SECTION

A70031

A70031

1/2 (12.7) DIA HOLE
THRU
2 REQ'D.

1 7/8
(47.6)

3 1/4
(82.4)

1 5/8
(41.2)

N

M

L

5/8 (16) DIA. HOLE
3/4 (19) DIA. C'BORE x 3/8 (9) DP.
2 REQ'D.

5/8
(15.8)

5/8
(15.8)

F

A

D

E

C

B

1 1/2
(38.1)

1 3/4
(44.4)

A

A

5/8
(15.8)

1 1/4
(31.7)

5/8
(15.8)

1 1/4
(31.7)

G

H

3/8 (9.5)

7/8
(22.2)

K

J

I

5.0
(127)

2 1/2
(63.5)

Section A-A

	DRAWN JCB	10/8/81	CHECKED NCB	10/10/81	INCH MILLIMETER
	APPROV. JCM	10/15/81	MATERIAL CAST IRON		

SUPPORT-BLOCK

JAN ENGINEERING
Box 52
LYNDONVILLE, VT.

DRAWING NO.
A70031

DO NOT SCALE DRAWING

TOLERANCE UNLESS
OTHERWISE SPEC.
.XX ± .015
.XXX ± .005
.XXXX ± .0002

HEAT TREATMENT

SCALE HALF SIZE

LET	CHANGE	DATE

CHANGE NOTICE

HALF SECTION

A half section is a drawing where the cutting line passes halfway through an object and a *quarter* section is removed as illustrated below. A half section allows you to view *both* the inside of the object *and* the outside of the object in the same view.

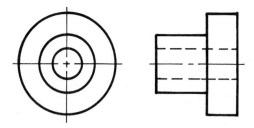

REGULAR 2-VIEW DRAWING

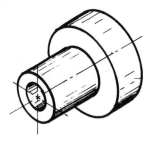

PICTORIAL VIEW

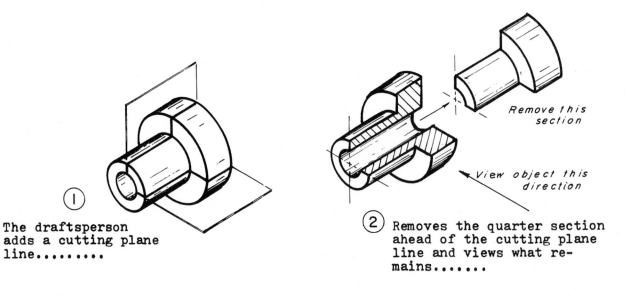

① The draftsperson adds a cutting plane line.........

② Removes the quarter section ahead of the cutting plane line and views what remains.......

Remove this section

View object this direction

③ Notice, the front view contains the cutting plane line.

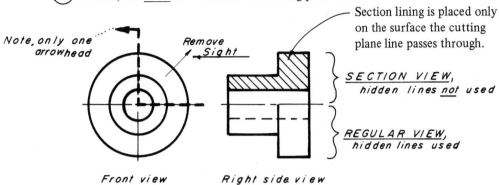

Note, only one arrowhead

Remove Sight

Section lining is placed only on the surface the cutting plane line passes through.

SECTION VIEW,
hidden lines not used

REGULAR VIEW,
hidden lines used

Front view   Right side view

RIGHT SIDE VIEW AS A HALF SECTION

HALF SECTION WORKSHEET

Using plan A8635-B, answer the following questions: (Note the location of the cutting plane line.) See end of chapter for answers.

1. Surface 'B' in the top view is what surface in the front view? _____

 _____

2. What is the diameter at 'C'? _____

3. What is the diameter at 'D'? _____

4. What is the radius at 'G'? _____

5. Calculate dimension 'I'. _____

6. What is the diameter of 'K'. _____

7. Surface 'L' in the front view is what surface in the top view? _____

 _____

8. What is the radius at 'N'? _____

9. Surface 'O' in the front view is what surface in the top view? _____

 _____

10. What was the size of the .75 (19) diameter when the drawing was issued?

 _____

11. What is the material? _____

12. Calculate dimension 'P'. _____

HALF SECTION VIEW

WORKSHEET

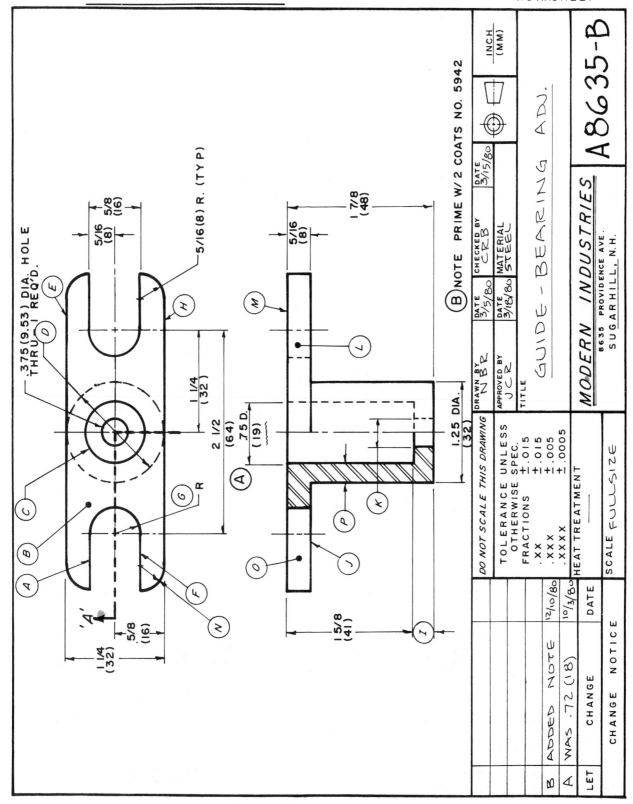

.375(9.53) DIA. HOLE
THRU—1 REQ'D.

5/16(8) R. (TYP)

5/8 (16)
5/16 (8)

1 1/4 (32)

2 1/2 (64)

.75 D. (19)

R

1 1/4 (32)

5/8 (16)

'A'

5/16 (8)

1 7/8 (48)

1.25 DIA. (32)

1 5/8 (41)

Ⓑ NOTE PRIME W/ 2 COATS NO. 5942

| DRAWN BY N B R | DATE 3/5/80 | CHECKED BY C R B | DATE 3/5/80 |
| APPROVED BY J C R | DATE 3/18/80 | MATERIAL STEEL | |

TITLE GUIDE—BEARING ADJ.

MODERN INDUSTRIES

8635 PROVIDENCE AVE.
SUGARHILL, N.H.

A8635-B

INCH (MM)

DO NOT SCALE THIS DRAWING

TOLERANCE UNLESS
OTHERWISE SPEC.
FRACTIONS ± .015
.XX ± .015
.XXX ± .005
.XXXX ± .0005

HEAT TREATMENT ———

SCALE FULLSIZE

			DATE
B	ADDED NOTE	12/10/80	
A	WAS .72 (18)	10/3/80	
LET	CHANGE		DATE

CHANGE NOTICE

BROKEN-OUT SECTION

Sometimes only a small area needs to be sectioned in order to make a drawing clear. In broken-out section drawings, the cutting plane is omitted.

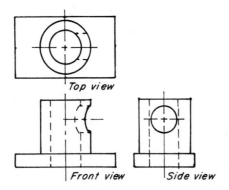

Top view

Front view   *Side view*

REGULAR 3-VIEW DRAWING

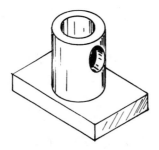

PICTORIAL VIEW

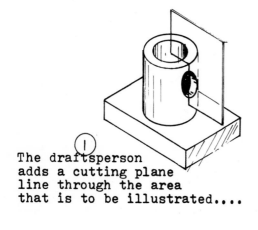

① The draftsperson adds a cutting plane line through the area that is to be illustrated....

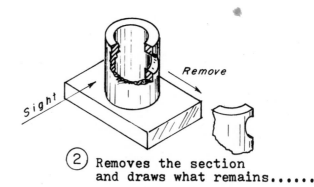

Sight   *Remove*

② Removes the section and draws what remains......

Converting back to a regular 3-view drawing, the broken-out area appears in the front view and illustrates the hole clearer.

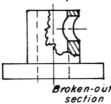

Top view

Broken-out
section

Side view

Section lining is added to area that has been cut out. Remember there is <u>no</u> cutting plane line in this type of a section view.

BROKEN-OUT SECTION WORKSHEET

Using plan A49371, answer the following questions: Notice, broken-out sections do *not* have a cutting plane line. See end of chapter for answers.

1. What is the dia. of hole '*A*'? _____

2. Surface '*B*' in the front view is what surface in the right side view? _____
_____

3. Surface '*D*' in the front view is what surface in the right side view? _____
_____

4. What are the upper and lower limits of hole '*E*'? _____

5. What is diameter '*L*'? _____

6. Calculate dimension '*G*'. _____

7. What is dia. '*H*'? _____

8. Surface '*I*' in the right side view is what lettered diameter in the front view?
_____

9. Surface '*J*' in the right side view is what surface in the front view? _____
_____

10. What is the dia. of diameter '*K*'? _____

11. What is the upper limit of the 2.250 (57.2) diameter? _____

12. What is the lower limit for the smallest hole in the bearing housing? _____
_____

BROKEN-OUT SECTION

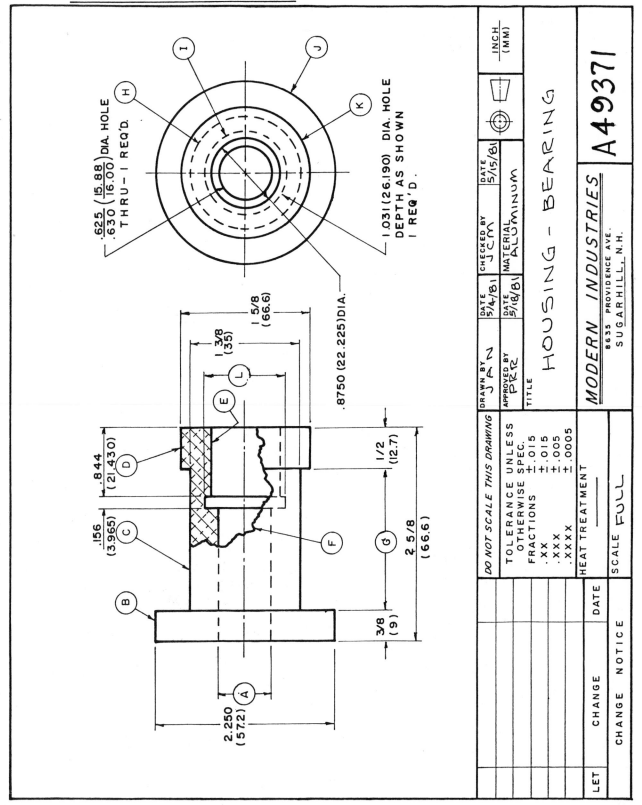

.625 (15.88) DIA. HOLE
.630 (16.00)
THRU—I REQ'D.

1.031 (26.190) DIA. HOLE
DEPTH AS SHOWN
I REQ'D.

.8750 (22.225) DIA.

1 5/8
(66.6)

1 3/8
(35)

.844
(21.430)

.156
(3.965)

2 5/8
(66.6)

1/2
(12.7)

3/8
(9)

2.250
(57.2)

INCH (MM)		
DRAWN BY J A N	DATE 5/4/81	CHECKED BY JCM	DATE 5/15/81

APPROVED BY P.R.R. DATE 5/18/81

MATERIAL ALUMINUM

TITLE

HOUSING – BEARING

MODERN INDUSTRIES
8635 PROVIDENCE AVE.
SUGARHILL, N.H.

A4937I

DO NOT SCALE THIS DRAWING

TOLERANCE UNLESS
OTHERWISE SPEC.
FRACTIONS ±.015
.XX ±.015
.XXX ±.005
.XXXX ±.0005

HEAT TREATMENT ———

SCALE FULL

LET | CHANGE | DATE

CHANGE NOTICE

OFFSET SECTION

In order to include as many important features of an object, the cutting plane line can be drawn with 90° bends. Thus, it will pass through many more important features than if it passed straight through the object like a full section. The 90° bends are *not* illustrated in the section view.

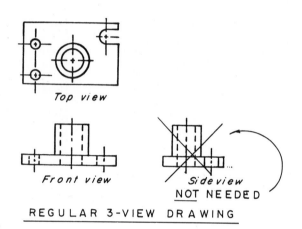

Top view

Front view   Side view
                          NOT NEEDED

REGULAR 3-VIEW DRAWING

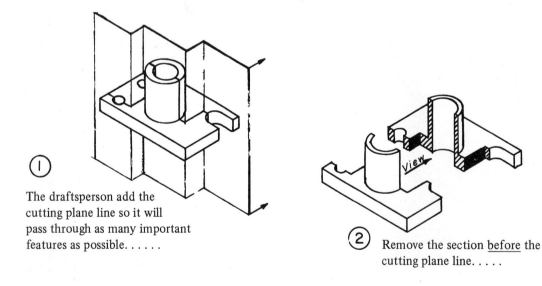

① The draftsperson add the cutting plane line so it will pass through as many important features as possible. . . . . .

② Remove the section before the cutting plane line. . . . .

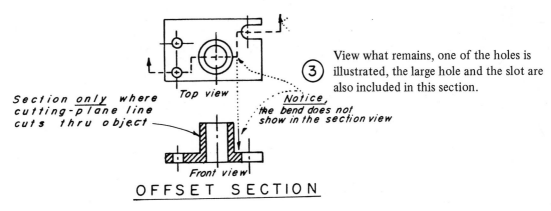

Section only where cutting-plane line cuts thru object

Top view

③ View what remains, one of the holes is illustrated, the large hole and the slot are also included in this section.

Notice, the bend does not show in the section view

Front view

OFFSET SECTION

Right side view no longer needed......

OFFSET SECTION WORKSHEET

Using plan #A843314, answer the following questions. See end of chapter for answers.

1. What is dimension 'A'? _____

2. What is dimension 'B'? _____

3. What is radius 'E'? _____

4. What is dimension 'H'? _____

5. What is dimension 'I'? _____

6. What is the overall length 'J'? _____

7. What is radius 'K'? _____

8. Surface 'L' in the front view is what surface in the top view? _____
    _____

9. Calculate dimension 'M'. _____

10. Surface 'N' in the right side view represents the top of what? _____
    _____

11. Surface 'O' in the right side view is what surface in the top view? _____
    _____

12. What is dimension 'P'? _____

13. What is dimension 'Q'? _____

14. The countersunk hole in *section A-A* is which hole in the top view 'a' or 'b'?
    _____

15. What is the distance from the C' bored hole to the C' sunk holes? _____
    _____

OFFSET SECTION

SECTION A-A

ALL UNMARKED RADIUS 1/16(2) R UNLESS OTHERWISE NOTED

1/2 (12.7) DIA. HOLE-THRU
13/16(20.6) C'BORE × 9/16 (14.2) DP.
1 REQ'D. 7/8 (22.2) DIA.(TYP.)

5/16 (7.9) DIA. HOLE-THRU
82° × 5/8 (15.8) DIA. C'SINK
2 REQ'D.

1/4 (6.3) R. (TYP.)

1 5/16
(33.7)

7/8
(22)

5/8
(16)

5/16
(8)

7/16
(11)

5/16
(8)

1.000
(25.40)

.500
(12.70)

2.00 (50.8) DIA.

2.125
(55.98)

3.750
(95.25)

5 1/2
(139.7)

1/8 R
(3)

1 1/2
(38.1)

11/16
(17.4)

5/8
(15.8)

1/2
(12.7)

$\frac{1}{8}$R
(3)

DO NOT SCALE THIS DRAWING

TOLERANCE UNLESS
OTHERWISE SPEC.
FRACTIONS ± .015
.XX ± .015
.XXX ± .005
.XXXX ± .0005

HEAT TREATMENT

SCALE 3/4 SIZE

DRAWN BY JCM | DATE 8/6/81
APPROVED BY JCM | DATE 8/8/81
CHECKED BY JAN | DATE 8/6/81
MATERIAL BRASS

TITLE
ARM-ADJUSTING

MODERN INDUSTRIES
8635 PROVIDENCE AVE.
SUGARHILL, N.H.

A843314

INCH
(MM)

LET | CHANGE | DATE
CHANGE NOTICE

FILLETS, ROUNDS, AND RUNOUTS

Usually, anything that is cast, has *"rounds"* or *"fillets"* included in its design. Rounds and fillets are used for three reasons:

1. Appearance,
2. Strength, and
3. To eliminate all sharp corners.

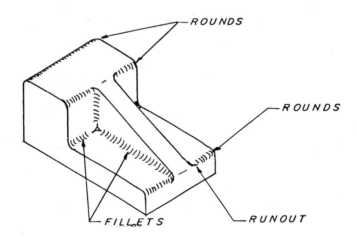

"Runout" is where one surface "runs" or blends into another surface.

Thin Wall Section. Any object that is drawn "in-section" and is very thin—such as a gasket or brass shim—is filled in solid black as it is impossible to show the correct section lining. This would be called a *Thin Wall Section.* If the cutting plane passes along the center of a bolt, screw, nut, pin, key, shaft or rivet, it is *not sectioned.* For example:

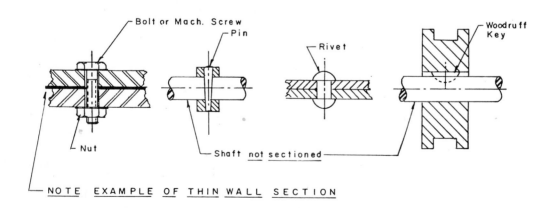

Note thin wall section

ROTATED SECTION

In some drawings, a portion of the object is not clear as to its exact shape or size. The draftsperson will *rotate* the unclear portion, *in its place,* thus showing its shape and size.

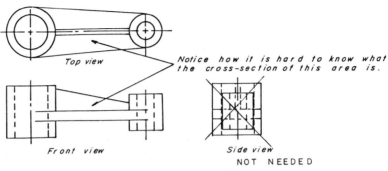

Top view

Notice how it is hard to know what the cross-section of this area is.

Front view

Side view
NOT NEEDED

REGULAR 3-VIEW DRAWING

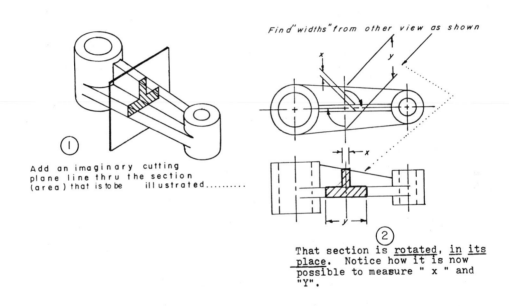

① Add an imaginary cutting plane line thru the section (area) that is to be illustrated..........

Find "widths" from other view as shown

② That section is <u>rotated</u>, <u>in its place</u>. Notice how it is now possible to measure " x " and "Y".

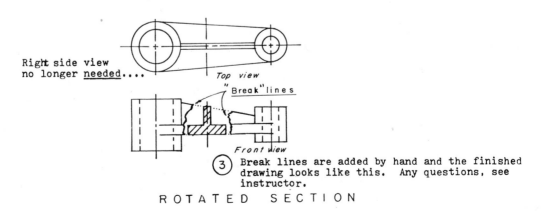

Right side view no longer <u>needed</u>....

Top view

"Break" lines

Front view

③ Break lines are added by hand and the finished drawing looks like this. Any questions, see instructor.

ROTATED SECTION

ROTATED SECTION WORKSHEET

Using plan #A693254 fill in all answers below. See end of chapter for answers.

1. What is diameter 'A'? _____

2. What is dimension 'B'? _____

3. What is radius 'E'? _____

4. Calculate dimension 'G'. _____

5. What is diameter 'H'? _____

6. Surface 'I' in the front view is what surface in the top view? _____
_____

7. What is dimension 'J'? _____

8. Surface 'K' in the front view is what surface in the top view? _____
_____

9. Diameter 'L' is what? _____

10. What is the line at 'M' called? _____

11. What is dimension 'N'? _____

12. What is dimension 'O'? _____

13. What is the area at 'P' called? _____

14. What scale is this drawing drawn to? _____

15. What material is used in this drawing? _____

ROTATED SECTION

A693254

2 1/4 (57.1) DIA.

5/16(7.9)

5/8 (15.8)

1 1/16 (26.9) DIA. HOLE THRU — 1 REQ'D.

7.250 (154.15)

3 5/8 (92)

2 3/16(55.5) DIA. HOLE THRU — 1 REQ'D.

2 13/16 (71.4)

1 13/32 (35.7)

3/8(9.5)

1/2 (12.7)

1 7/16 (36.5)

2 3/4 (69.9)

3/4 (19)

2.00 (50.8)

ALL FILLETS/ROUNDS 1/8 (3.1) R

	INCH
	MILLIMETER

ARM - CENTER

DRAWN	JAD	7/4/80	CHECKED	RGP	7/10/80
APPROV.	JCM	7/15/80	MATERIAL	CAST IRON	

JAN ENGINEERING
Box 52
LYNDONVILLE, VT.

DRAWING NO.
A693254

DO NOT SCALE DRAWING

TOLERANCE UNLESS OTHERWISE SPEC
.XX ± .015
.XXX ± .005
.XXXX ± .0002

HEAT TREATMENT

SCALE HALF SIZE

LET	CHANGE	DATE
CHANGE NOTICE		

REMOVED SECTION

A removed section is the same as a revolved section *except* as the name implies, the section is *"removed"* and drawn away from the object. Study the examples below—note how they are "called off" (i.e., Section A-A, Section B-B, etc.) Compare this rotated section, with the removed section.

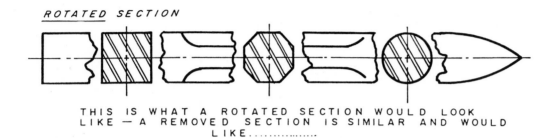

ROTATED SECTION

THIS IS WHAT A ROTATED SECTION WOULD LOOK
LIKE — A REMOVED SECTION IS SIMILAR AND WOULD
LIKE.................

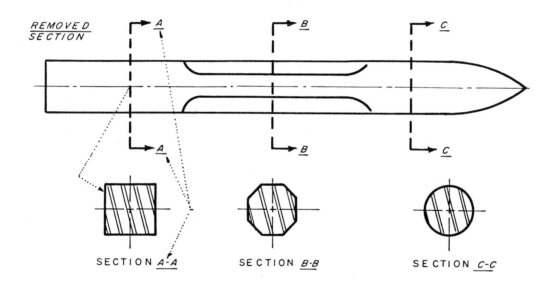

REMOVED SECTION

SECTION A-A   SECTION B-B   SECTION C-C

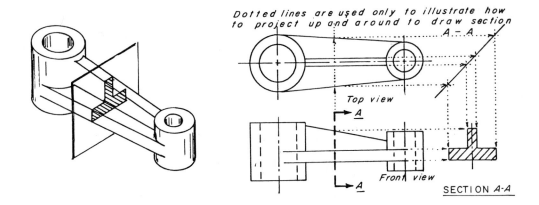

Refer back to the *rotated* section, compare it with this *removed section*. All three steps are the same, except the section area is *removed*.

REMOVED SECTION WORKSHEET

Using drawing A769325 answer the following questions. See end of chapter for answers.

1. What is dimension 'A'? _____

2. What is dimension 'E'? _____

3. Surface 'H' in the left side view is what surface in the top view? _____
    _____

4. What is dimension 'I'? _____

5. What is dimension 'J'? _____

6. Surface 'K' in the front view is what surface in the left side view? _____
    _____

7. What is dimension 'L'? _____

8. Surface 'M' in the front view is what surface in the right side view? _____
    _____

9. What is dimension 'N'? _____

10. Calculate dimension 'O'. _____

11. Surface 'P' in the right hand view is what surface in the top view? _____
    _____

12. Surface 'R' in the right hand view is what surface in the top view? _____
    _____

13. What is diameter 'S'? _____

14. What is diameter 'T'? _____

15. What is dimension 'U'? _____

16. What is diameter 'V'? _____

17. What is dimension 'W'? _____

18. Surface 'X' in Section B-B is what surface in the front view? _____

19. How deep does 'Y' diameter go into the part? _____

20. Surface 'Z' in Section B-B is what surface in the top view? _____

REMOVED SECTION   WORKSHEET

SECTION B-B

1.00 (25.4) DIA.

.688 (17.4)

1.375 (34.9)

SECTION A-A

45°

90°

2.062 (52.38)

.812 (20.640) DIA.

11/16 (17.4)

3/32 (2.3)

.625 (15.88) DIA. HOLE — THRU TO CTR HOLE
.875 (22.2) C'BORE × .125 (3.2) DEEP
1 REQ'D

.25 (6.8)

7/8 (22.2)

4.00 (101.6)

2 11/16 (68.27)

1.000 (25.40)

.500 (12.70)

2 3/8 (60.3)

3/4 (19)

1.500 (38.1)

.625 (15.88) DIA. HOLE
THRU — 1 REQ'D

1 3/4 ACROSS FLATS
(44.4)

7/8 (22.2)

DO NOT SCALE THIS DRAWING

DRAWN BY RCB DATE 10/5/80
APPROVED BY JAN DATE 10/8/80
CHECKED BY RCB DATE 10/6/80

MATERIAL STEEL 2.062 D × 5½ LG (5238)M (139.7)

TITLE
BLOCK — ADJUSTING

MODERN INDUSTRIES
8635 PROVIDENCE AVE.
SUGARHILL, N.H.

A769325

INCH (MM)

TOLERANCE UNLESS
OTHERWISE SPEC.
FRACTIONS ± .015
.XX ± .015
.XXX ± .005
.XXXX ± .0005

HEAT TREATMENT

SCALE 3/4 SIZE

LET   CHANGE   DATE

CHANGE NOTICE

ASSEMBLY SECTION

An assembly section illustrates the parts of an assembly much clearer than the conventional two- or three-view drawing. At the left is a two-view drawing of an assembly. Note the right side view is hard to understand. To the right is a two-view drawing with the right side via a full section, notice how it is much easier to understand.

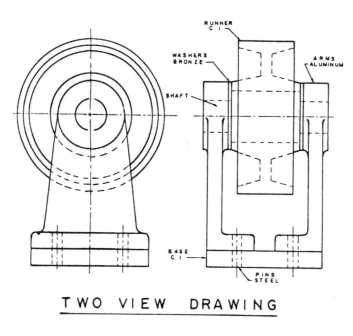

TWO VIEW DRAWING

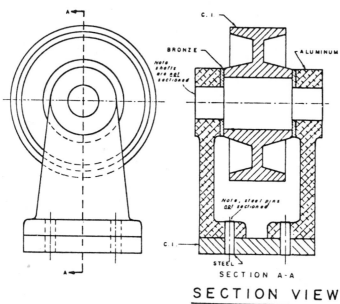

SECTION A-A

SECTION VIEW

<u>Review</u>: If the cutting plane line passes through the center of a bolt, screw, nut, key, shaft, rivet, it is <u>not</u> sectioned.

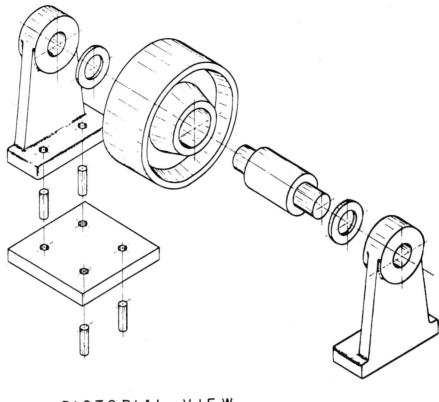

PICTORIAL VIEW

As a rule, assembly <u>sections</u> do <u>not</u> have dimensions. (Each part is dimensioned on its own drawing).

In the pictorial view of the above section view, be sure you fully understand the section view, its various parts, their locations, and the material used on making each before proceeding.

ASSEMBLY SECTION WORKSHEET

Using drawing A4112795, answer the following questions in the space provided. (Study this drawing very carefully before answering the questions, be sure you have a mental picture of what each part looks like. See end of chapter for answers.

1. How many parts are there in this assembly section drawing (not counting the lead)? _____

2. What kind of a section view is the left side of the compass (parts 4, 3, 11, 10)? _____

3. What is the material used in this assembly section? _____

4. Section *A-A* is what kind of a section view? _____

5. What parts are used more than once? _____

6. What is the scale used to draw this section? _____

7. Why are dimensions omitted? _____

8. Describe the shape of part no. 1. _____

9. What part numbers hold the lead? _____

10. How many days did it take from the time the drawing was checked until it was approved? _____

ASSEMBLY SECTION

Section A-A

INCH
(MM)

DATE 5/8/81

CHECKED BY
RCB

DATE 5/8/81

MATERIAL
AS NOTED

DRAWN BY
DRB

DATE 5/8/81

APPROVED BY
JAN

DATE 5/9/81

TITLE

COMPASS-BEAM

MODERN INDUSTRIES

8635 PROVIDENCE AVE.
SUGARHILL, N.H.

A 4112795

DO NOT SCALE THIS DRAWING

TOLERANCE UNLESS
OTHERWISE SPEC.
FRACTIONS ± .015
.XX   ± .015
.XXX   ± .005
.XXXX   ± .0005

HEAT TREATMENT

SCALE NONE

LET | CHANGE | DATE

CHANGE NOTICE

CONVENTIONAL DRAFTING PRACTICES

Sometimes a drawing is actually illustrated incorrectly in order to better illustrate a part. These incorrect drafting practices have general rules, a few are illustrated in the next few pages.

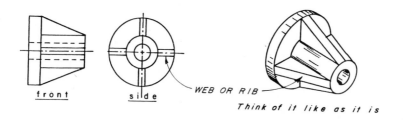

front   side   *WEB OR RIB*

Think of it like as it is

NOW, ADD A REGULAR CUTTING PLANE LINE TO THE SIDE................

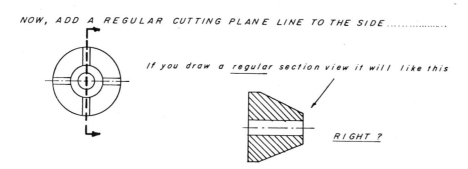

If you draw a regular section view it will like this

RIGHT ?

THAT MAKE IT APPEAR TO LOOK LIKE.........................

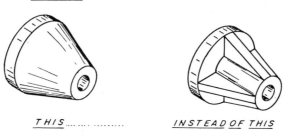

THIS..................   *INSTEAD OF THIS*

THUS THE RULE:
DO NOT SECTION-LINE RIBS OR WEBS

Note web or rib not lined

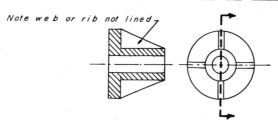

CONVENTIONAL METHOD

CONVENTIONAL PRACTICES: RIBS AND WEBS

A rib or web is an area that holds other major parts together. In this case, a rib is drawn incorrectly in order to better illustrate the part.

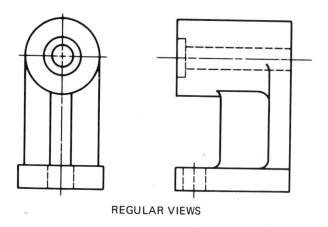

REGULAR VIEWS

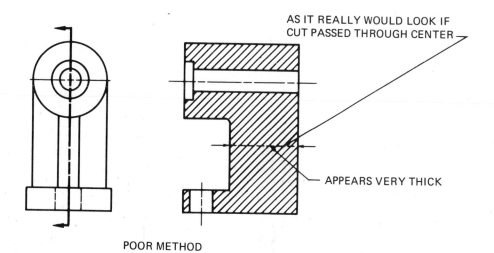

AS IT REALLY WOULD LOOK IF
CUT PASSED THROUGH CENTER

APPEARS VERY THICK

POOR METHOD

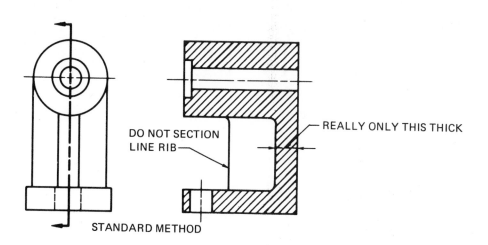

REALLY ONLY THIS THICK

DO NOT SECTION
LINE RIB

STANDARD METHOD

CONVENTIONAL PRACTICES: HOLE LOCATIONS

Hole locations are also drawn incorrectly in order to better illustrate the part and create less confusion. Study each drawing below. The left illustration is drawn correctly but is confusing and hard to understand. To the right is the conventional drafting method used to illustrate the part.

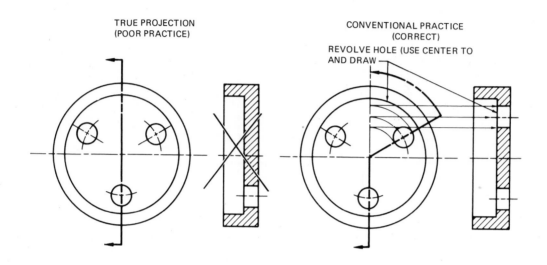

TRUE PROJECTION
(POOR PRACTICE)

CONVENTIONAL PRACTICE
(CORRECT)

REVOLVE HOLE (USE CENTER TO
AND DRAW

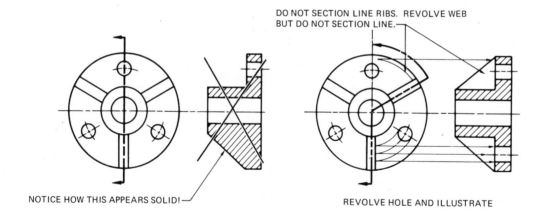

DO NOT SECTION LINE RIBS. REVOLVE WEB
BUT DO NOT SECTION LINE.

NOTICE HOW THIS APPEARS SOLID!

REVOLVE HOLE AND ILLUSTRATE

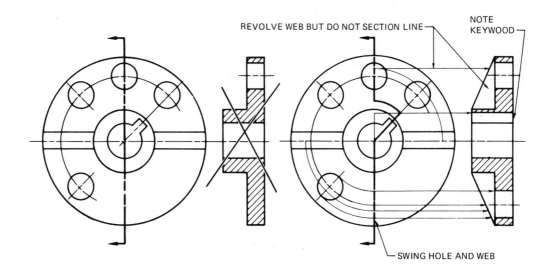

CONVENTIONAL PRACTICES WORKSHEET

Using drawing A49372-A, answer the following questions in the space provided. See end of chapter for answers.

1. What is radius 'A'? _____

2. What is diameter 'B'? _____

3. What is the lower limit of hole 'C'? _____

4. What is the upper limit of hole 'D'? _____

5. What is the diameter of hole 'E'? _____

6. What is the radius at 'F' called? _____

7. Calculate dimension 'G'. _____

8. What is radius 'H' called? _____

9. The web at 'I' is not sectioned, why? _____

10. What is the radius at 'J' called? _____

11. What is dimension 'K'? _____

12. The symbol at 'L' indicates what? _____

13. What is the line at 'M' called? _____

14. The hole at 'N' is drawn in the incorrect location, as the cutting plane line does *not* pass through it, why is this? _____
_____

15. What was the *depth* of the 1.750 (44.45) dia. hole when the drawing was approved? _____

CONVENTIONAL PRACTICES

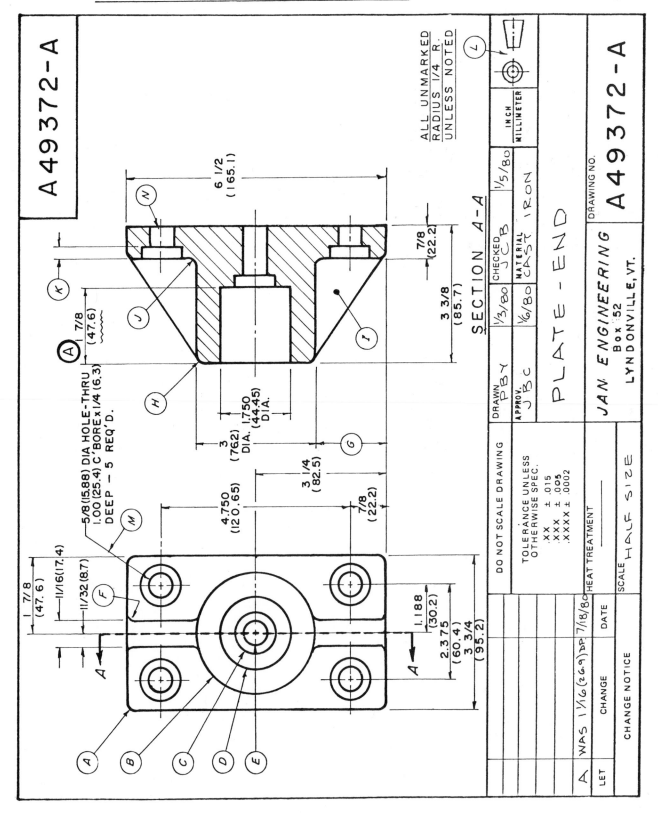

A49372-A

ALL UNMARKED
RADIUS 1/4 R.
UNLESS NOTED

SECTION A-A

| DRAWN PBY | 1/3/80 | CHECKED JCB | 1/5/80 | INCH MILLIMETER |
| APPROV. JBC | 1/6/80 | MATERIAL CAST IRON | | |

PLATE-END

JAN. ENGINEERING
Box 52
LYNDONVILLE, VT.

DRAWING NO.
A49372-A

DO NOT SCALE DRAWING

TOLERANCE UNLESS
OTHERWISE SPEC.
.XX ± .015
.XXX ± .005
.XXXX ± .0002

HEAT TREATMENT

SCALE HALF SIZE

| A | WAS 1 1/16 (26.9) DP 7/8/80 |
| LET | CHANGE | DATE |

CHANGE NOTICE

6 1/2
(165.1)

7/8
(22.2)

3 3/8
(85.7)

1 7/8
(47.6)

5/8 (15.88) DIA HOLE-THRU
1.00 (25.4) C'BORE x 1/4 (6.3)
DEEP – 5 REQ'D.

1.750
(44.45)
DIA.

3
(76.2)
DIA.

3 1/4
(82.5)

7/8
(22.2)

4.750
(120.65)

1 7/8
(47.6)

11/16 (17.4)

11/32 (8.7)

1.188
(30.2)

2.375
(60.4)

3 3/4
(95.2)

CONVENTIONAL PRACTICES: ROTATING

In drawing spokes, the draftsperson "rotates" the spoke(s) in line with the cutting plane line and draws them in that position.

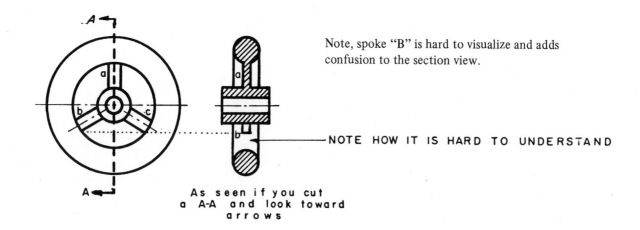

Note, spoke "B" is hard to visualize and adds confusion to the section view.

—NOTE HOW IT IS HARD TO UNDERSTAND

As seen if you cut
a A-A and look toward
arrows

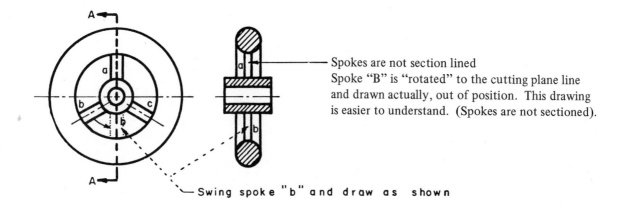

Spokes are not section lined
Spoke "B" is "rotated" to the cutting plane line and drawn actually, out of position. This drawing is easier to understand. (Spokes are not sectioned).

Swing spoke "b" and draw as shown

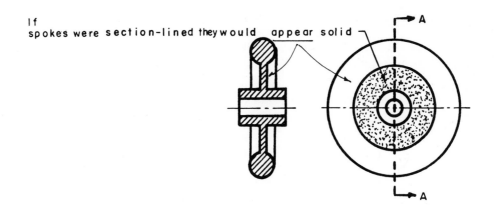

Remember, conventional drafting practices are actual methods used by the drafts-person to better illustrate a part without question and less confusion.

CONVENTION PRACTICES WORKSHEET

Using drawing A549923 answer the following questions in the spaces provided. See end of chapter for answers.

1. Spoke *'B'* in the front view is what surface in section A-A? _____
_____

2. Spoke *'G'* in the front view is what surface in section A-A? _____
_____

3. Calculate dimension *'I'*. _____

4. What is dimension *'J'*? _____

5. Surface *'K'* in section A-A is what surface in the front view? _____
_____

6. Hole *'L'* in section A-A is what surface in the front view? _____
_____

7. Spoke *'M'* in section A-A is which spoke in the front view? _____
_____

8. Calculate dimension *'N'*. _____

9. Hole *'P'* in section A-A is what hole in the front view? _____

10. What is radius *'Q'*? _____

11. What is dimension *'R'*? _____

12. What is dimension *'S'*? _____

13. How far apart are the .375 (.952) dia. holes? _____

14. What is the approximate size of the keyway? _____

CONVENTIONAL PRACTICES

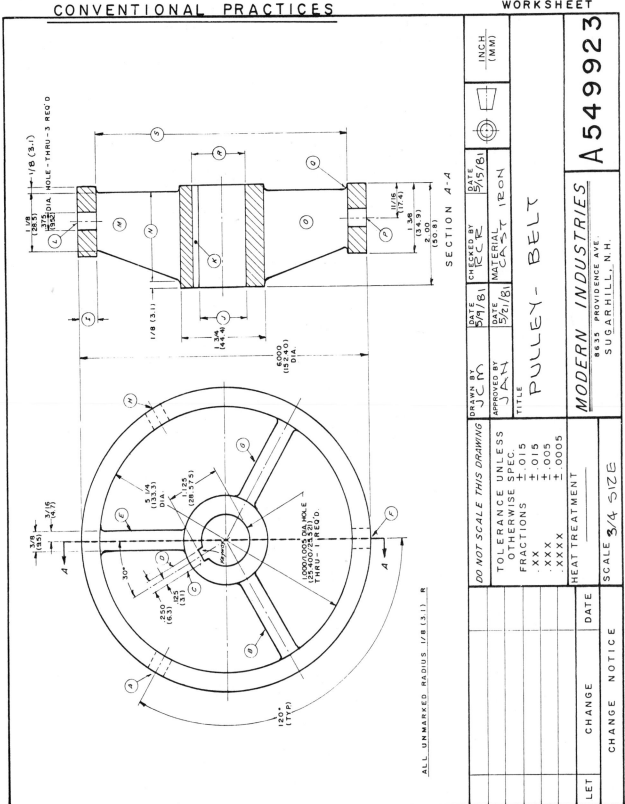

SECTION A-A

ALL UNMARKED RADIUS 1/8 (3.1) R

DO NOT SCALE THIS DRAWING	DRAWN BY JCM	APPROVED BY JAN

TITLE

PULLEY- BELT

MODERN INDUSTRIES
8635 PROVIDENCE AVE.
SUGARHILL, N.H.

CHECKED BY RCR	DATE 5/15/81
MATERIAL CAST IRON	

DATE 5/9/81	DATE 5/21/81

A549923

INCH (MM)

TOLERANCE UNLESS
OTHERWISE SPEC.
FRACTIONS ± .015
.XX ± .015
.XXX ± .005
.XXXX ± .0005

HEAT TREATMENT

SCALE 3/4 SIZE

LET	CHANGE	DATE

CHANGE NOTICE

CONVENTIONAL PRACTICES: BREAK LINES

In drawing a long sold square, rectangular, round pipe, tube or large flat plate, it is not necessary to show its complete length. It is usually drawn full size or a larger scale with a break line as illustrated below.

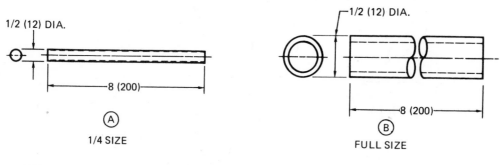

Use of conventional break

Example "A" is drawn 1/4 size in order to draw its *full* length, notice how hard it is to read. Example "B" was drawn *full size,* with a break line, and is much easier to understand.

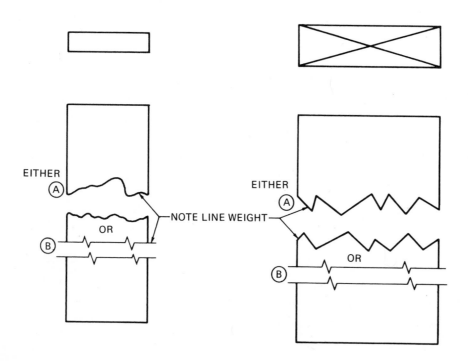

EVALUATION

–Answer all questions in the space provided:

1. When the cutting plane line passes through the center of some objects, they are NOT sectioned. List at least eight such objects that are NOT sectioned.

_____

_____

2. Are hidden lines usually used in a section view? _____

_____

3. Would a non-symmetrical object be drawn in half section? _____

4. Refer to question 3, why would you or why wouldn't you? _____

_____

5. At what angle is section lining usually drawn? _____

6. What is the "standard all purpose" section lining? _____

7. What kind of a section view shows the inside AND the outside of the object?

_____

8. On a cutting plane line, what do the arrows at each end mean? _____

_____

_____

9. When is a "conventional break" used and why? _____

_____

_____

10. Explain in full (use sketches if necessary) why ribs and spokes are NOT sectioned. _____

_____

_____

11. What kind of a section view illustrates only one area or one important feature? _____

_____

12. What kind of a section view is similar to a full section *except* that the cutting plane line bends? Why is it drawn with 90° bends? _____

_____

_____

13. List three reasons 'rounds' and 'fillets' are used? _____

_____

14. What kind of a section view is similar to a rotated section? _____

_____

15. What kind of section view illustrates how various parts are assembled. ____

_____

16. Why are *ribs* and *webs* not sectioned? _____
_____
_____
_____

17. Why are holes sometimes located incorrectly? _____
_____
_____

18. Why are spokes *not* sectioned? _____
_____
_____

ANSWERS TO TERMS TO KNOW AND UNDERSTAND

Cutting Plane Line—An imaginary plane passed through the object. The cutting plane line shows where the object will be viewed. (Think of it as a saw cutting the object in two). It is a *thick* line.

Section Lining—The surface exposed by the cutting plane line is illustrated by very thin lines drawn at an angle. *These lines are located only on the surface where the cutting plane line touches.*

Symbols Used To Represent Various Materials—Section lining.

Kinds Of Sections:

Full Section—The cutting plane passed fully through the object, cutting it in two pieces.

Half Section—A section shown through one-half the object. The object should be symmetrical (same on both sides). The advantage of a half section is that it illustrates both the exterior of the object (one side) and the interior of the object (the other side).

Broken-out Section—A partial interior view is done by a broken-out section. The cutting plane line is *not* drawn.

Offset Section—Very much like a full section except the cutting plane line changes direction in order to show various important features.

Rotated Section—(Sometimes called revolved). Is used to illustrate the cross-section of an object, and is drawn in place, rotated 90°.

Removed Section—Very much like a rotated section, except the area to be illustrated is rotated and drawn in a convenient location on the drawing *away* from the object. This method is used more than a rotated section.

Assembly Section—Used to illustrate how a device is assembled and sometimes how it functions. Each part is cross-sectioned a different direction in order to clearly illustrate each part.

Conventional Breaks—Used in the event an object cannot fit on the drawing paper. The object is shortened by using break lines.

'Fillets'—An area *filled* in by a radii.

'Rounds'—An edge *rounded-off* by a radii.

'Run-outs'—Where one surface *fades* or *blends* into another surface.

FULL SECTION WORKSHEET ANSWERS

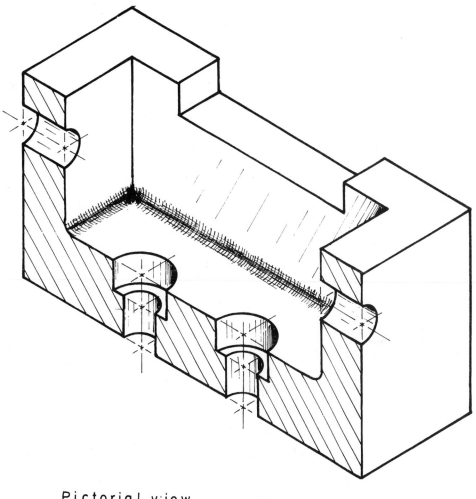

Pictorial view

A70031

FULL SECTION WORKSHEET ANSWERS

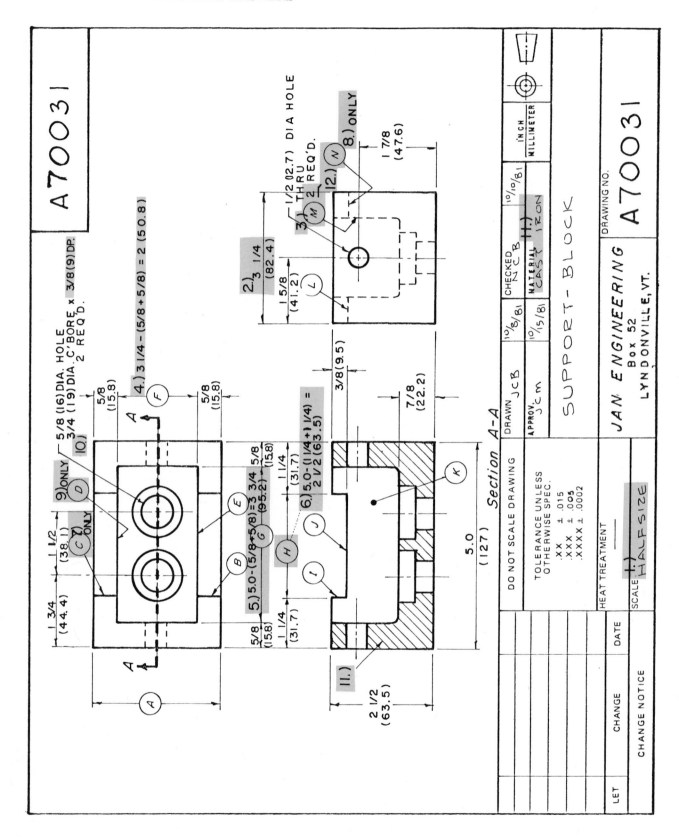

A70031

5/8 (16) DIA. HOLE
3/4 (19) DIA. C'BORE x 3/8 (9) DP.
2 REQ'D.

10.)

4.) 3 1/4 − (5/8 + 5/8) = 2 (50.8)

5/8 (15.8)

F

5/8 (15.8)

A

9) ONLY

D

7) ONLY

C

1 3/4 (44.4)

1 1/2 (38.1)

E

B

5/8 (15.8)

5.) 5.0 − (5/8 + 5/8) = 3 3/4 (95.2)

G

5/8 (15.8)

1 1/4 (31.7)

6.) 5.0 − (1 1/4 + 1 1/4) = 2 1/2 (63.5)

H

J

I

5/8 (15.8)

1 1/4 (31.7)

A

A

1/2 (12.7) DIA HOLE
THRU
2 REQ'D.

3.)

M

2.) 3 1/4 (82.4)

1 5/8 (41.2)

L

12.)

N

8.) ONLY

1 7/8 (47.6)

3/8 (9.5)

7/8 (22.2)

K

5.0 (127)

11.)

2 1/2 (63.5)

Section A-A

DRAWN JCB	10/8/81

SUPPORT - BLOCK

CHECKED N.C.B	10/10/81

APPROV. JCM	10/15/81

MATERIAL CAST IRON   7.)

JAN ENGINEERING
Box 52
LYNDONVILLE, VT.

DRAWING NO.

A70031

INCH MILLIMETER

DO NOT SCALE DRAWING

TOLERANCE UNLESS
OTHERWISE SPEC.
.XX ± .015
.XXX ± .005
.XXXX ± .0002

HEAT TREATMENT

SCALE HALFSIZE   1.)

LET	CHANGE	DATE

CHANGE NOTICE

HALF SECTION WORKSHEET ANSWERS

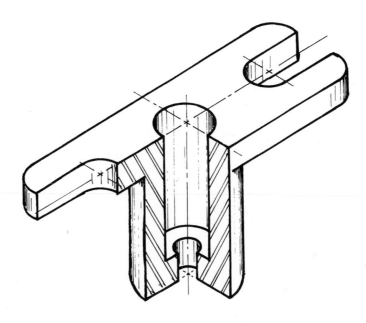

Pictorial view

HALF SECTION WORKSHEET ANSWERS

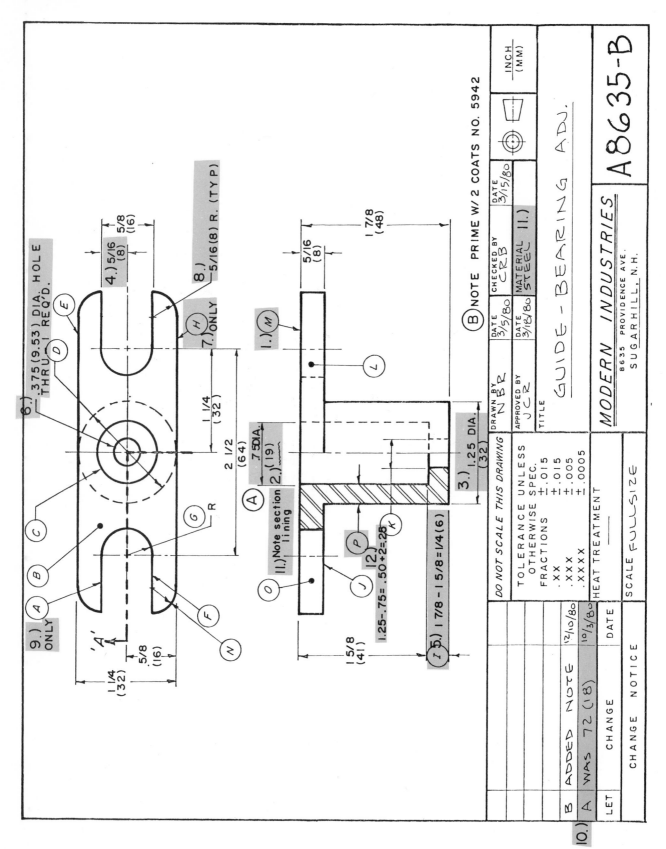

6.) .375 (9.53) DIA. HOLE THRU—1 REQ'D.

4.) 5/16 (8)   5/8 (16)

8.) 5/16 (8) R. (TYP)

7.) ONLY   (H)

1.) (M)

1 7/8 (48)

5/16 (8)

(L)

3.) 1.25 DIA. (32)

.75 DIA. (19)   2.)

(A)

11.) Note section lining

2 1/2 (64)

1 1/4 (32)

R   (G)

(C)   (B)   (D)   (E)

9.) (A) ONLY

'A'

1 1/4 (32)   5/8 (16)

(F)   (N)

(O)

1 5/8 (41)

(P)   12.)   (K)

(J)

1.25 − .75 = .50 ÷ 2 = .25

15.) 1 7/8 − 1 5/8 = 1/4 (6)   (I)

DO NOT SCALE THIS DRAWING

TOLERANCE UNLESS OTHERWISE SPEC.
FRACTIONS ±.015
.XX   ±.015
.XXX ±.005
.XXXX ±.0005

HEAT TREATMENT _____

SCALE FULLSIZE

DRAWN BY NBR
APPROVED BY JCR

CHECKED BY	DATE
CRB	3/5/80

| | DATE 3/5/80 |
| MATERIAL STEEL | DATE 3/18/80 |

11.)

(B) NOTE PRIME W/ 2 COATS NO. 5942

INCH (MM)

A8635-B

TITLE
GUIDE−BEARING ADJ.

MODERN INDUSTRIES
8635 PROVIDENCE AVE.
SUGAR HILL, N.H.

B	ADDED NOTE	12/10/80
A	WAS 72 (18)	10/13/80
LET	CHANGE	DATE

CHANGE NOTICE

10.)

BROKEN-OUT SECTION WORKSHEET ANSWERS

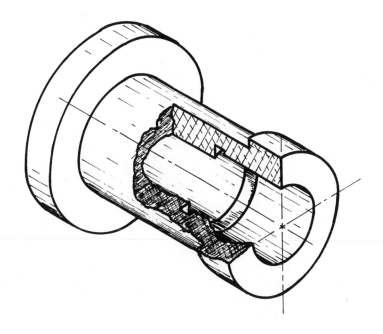

Pictorial view

A49371

BROKEN-OUT SECTION WORKSHEET ANSWERS

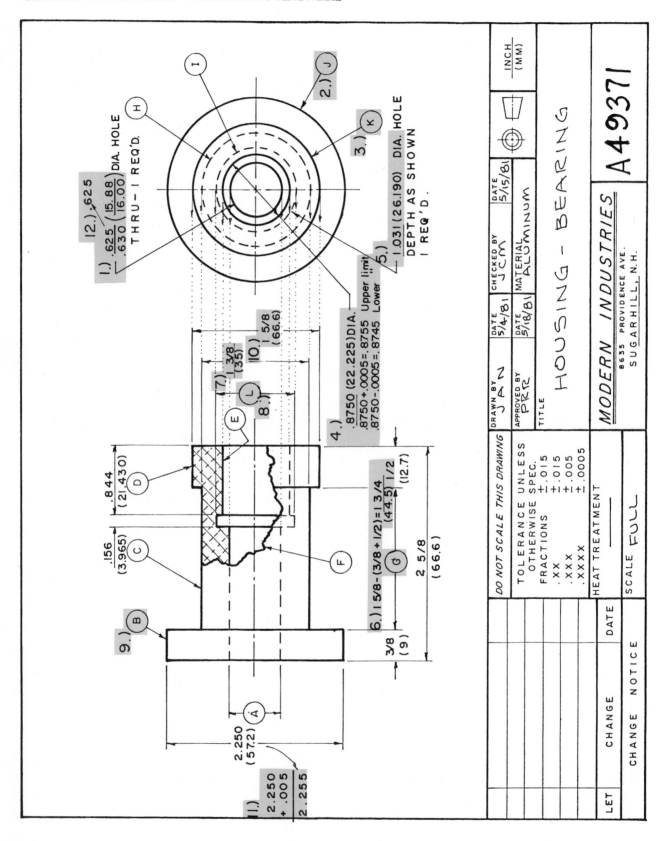

12.) .625
1.) $\frac{.625}{.630}\left(\frac{15.88}{16.00}\right)$ DIA. HOLE
THRU – 1 REQ'D.

3.) 1.031 (26.190) DIA. HOLE
DEPTH AS SHOWN
1 REQ'D.

.8750 (22.225) DIA.
.8750 + .0005 = .8755 Upper limit
4.) .8750 – .0005 = .8745 Lower

5.)

7.) $\frac{3/8}{(35)}$

10.) $\frac{5/8}{(66.6)}$

8.)

.844
(21.430)

.156
(3.965)

6.) 1 5/8 – (3/8 + 1/2) = 1 3/4
$\frac{(44.5)}{}$ 1/2
(12.7)

2 5/8
(66.6)

3/8
(9)

9.)

2.250
(57.2)

11.) 2.250
+ .005
2.255

DO NOT SCALE THIS DRAWING	DRAWN BY J A N	DATE 5/4/81	CHECKED BY JCM	DATE 5/15/81	INCH (MM)
	APPROVED BY PRR	DATE 5/18/81	MATERIAL ALUMINUM		

TOLERANCE UNLESS
OTHERWISE SPEC.
FRACTIONS ± .015
.XX ± .015
.XXX ± .005
.XXXX ± .0005

HEAT TREATMENT

SCALE FULL

TITLE

HOUSING – BEARING

MODERN INDUSTRIES
8635 PROVIDENCE AVE.
SUGARHILL, N.H.

A49371

LET	CHANGE	DATE

CHANGE NOTICE

OFFSET SECTION WORKSHEET ANSWERS

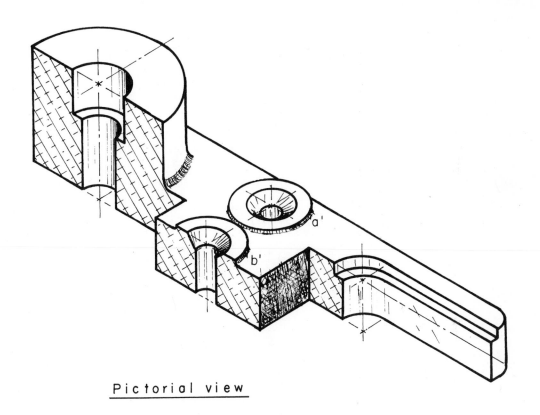

Pictorial view

A 843314

OFFSET SECTION WORKSHEET ANSWERS

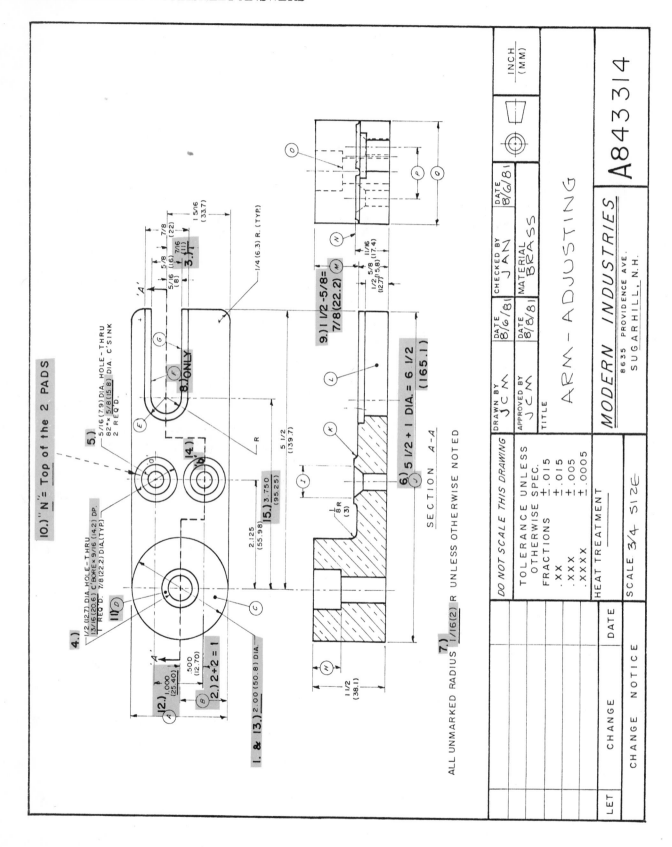

REVOLVED SECTION WORKSHEET ANSWERS

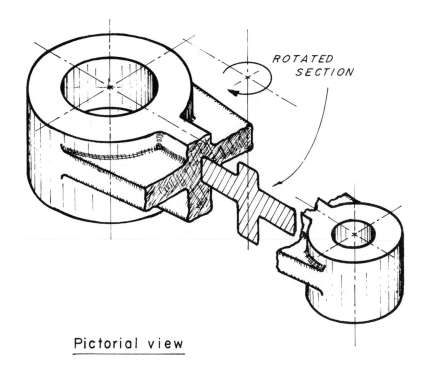

ROTATED SECTION

Pictorial view

A693254

ROTATED SECTION WORKSHEET ANSWERS

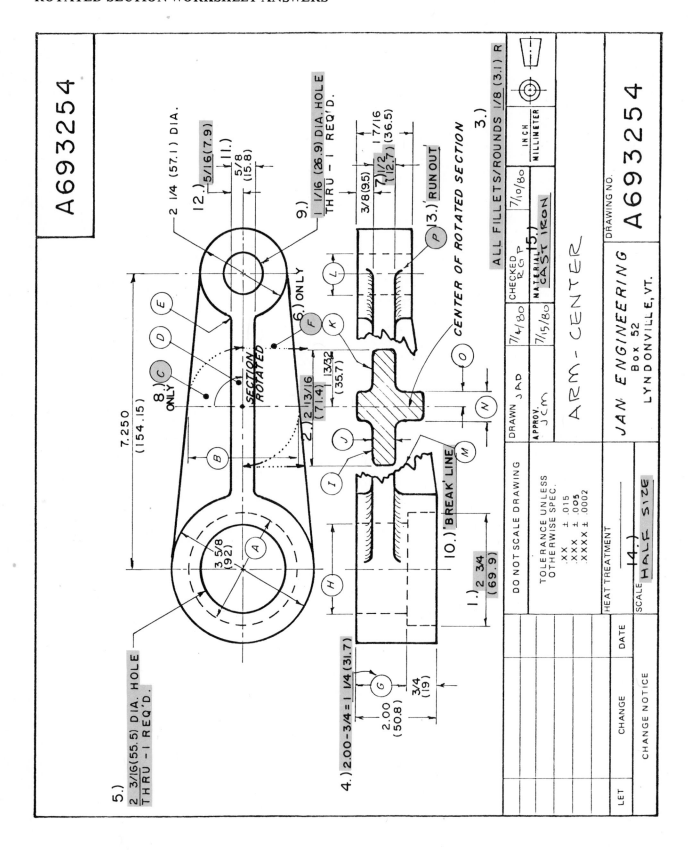

REMOVED SECTION WORKSHEET ANSWERS

REMOVED SECTION

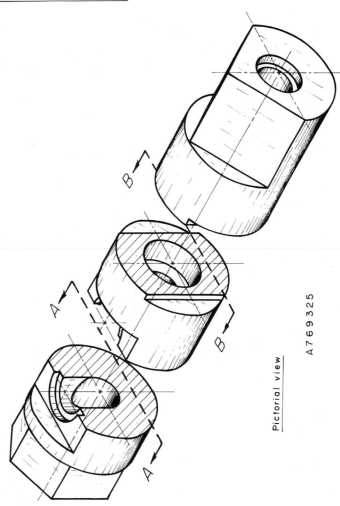

A769325

Pictorial view

REMOVED SECTION WORKSHEET ANSWERS

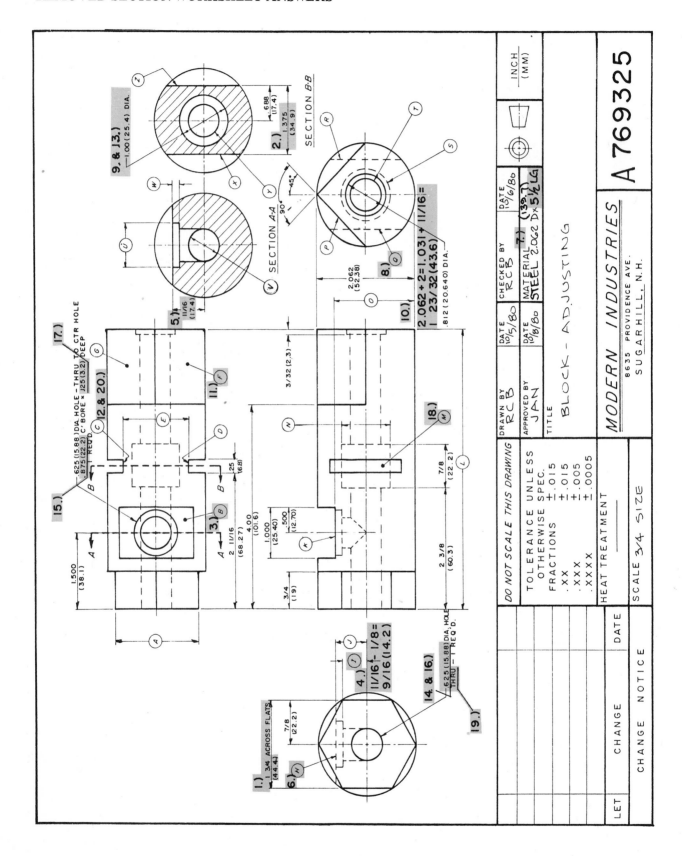

ASSEMBLY SECTION WORKSHEET ANSWERS

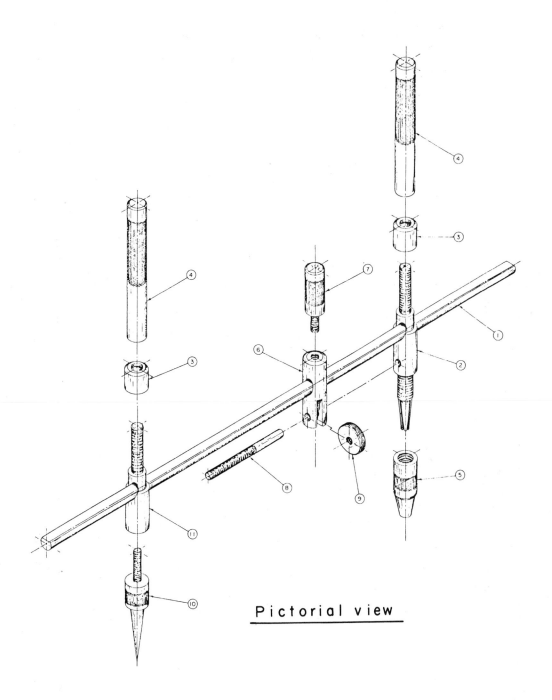

Pictorial view

A 4112795

ASSEMBLY SECTION WORKSHEET ANSWERS

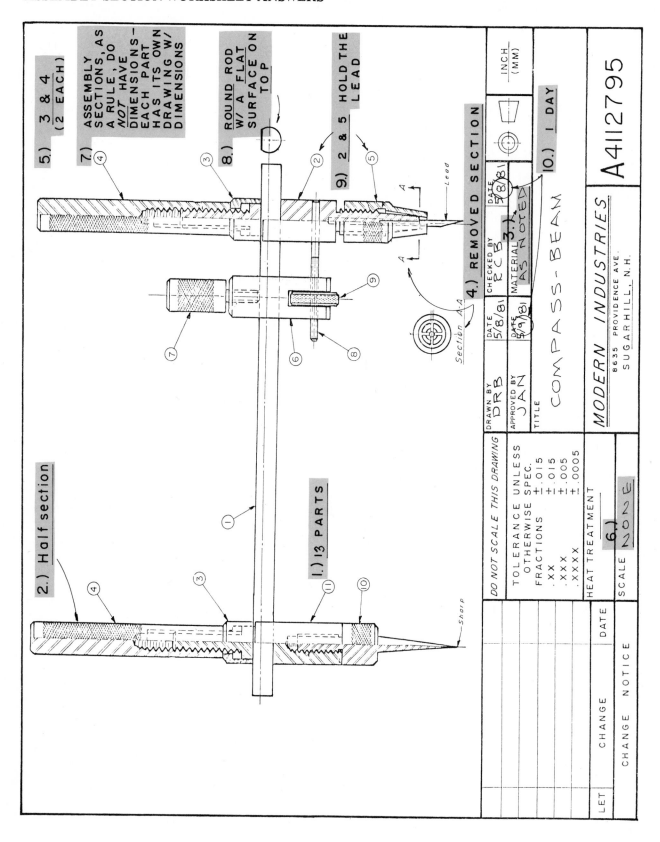

5.) 3 & 4 (2 EACH)

7.) ASSEMBLY SECTIONS, AS A RULE, DO *NOT* HAVE DIMENSIONS— EACH PART HAS ITS OWN DRAWING W/ DIMENSIONS

8.) ROUND ROD W/ A FLAT SURFACE ON TOP

9.) 2 & 5 HOLD THE LEAD

4.) REMOVED SECTION

2.) Half section

1.) 13 PARTS

Section A-A

Lead

Sharp

DRAWN BY DRB	DATE 5/8/81		INCH (MM)
APPROVED BY JAN	DATE 5/9/81	CHECKED BY RCB	DATE 5/8/81
		MATERIAL 3.) AS NOTED	10.) 1 DAY

TITLE COMPASS-BEAM

A4112795

MODERN INDUSTRIES
8635 PROVIDENCE AVE.
SUGARHILL, N.H.

DO NOT SCALE THIS DRAWING

TOLERANCE UNLESS OTHERWISE SPEC.
FRACTIONS ±.015
.XX ±1 ±.015
.XXX ±1 ±.005
.XXXX ±1 ±.0005

HEAT TREATMENT

6.) SCALE NONE

LET	CHANGE	DATE
	CHANGE NOTICE	

CONVENTIONAL PRACTICES WORKSHEET ANSWERS

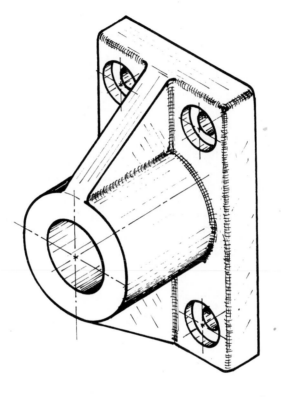

Pictorial view

A49372-A

CONVENTIONAL PRACTICES WORKSHEET ANSWERS

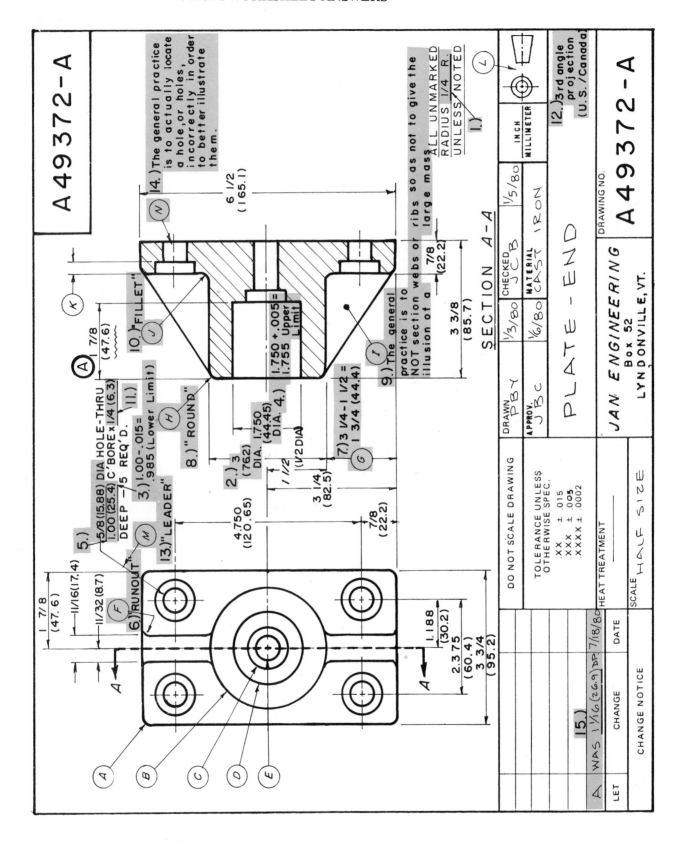

A49372-A

14.) The general practice is to actually locate a hole, or holes, incorrectly in order to better illustrate them.

1.) ALL UNMARKED RADIUS 1/4 R. UNLESS NOTED

9.) The general practice is to NOT section webs or ribs so as not to give the illusion of a large mass.

6 1/2 (165.1)

7/8 (22.2)

3 3/8 (85.7)

SECTION A-A

1.750 +.005 = 1.755 Upper Limit

10.) "FILLET"

1 7/8 (47.6)

5/8 (15.88) DIA HOLE-THRU 1.00 (25.4) C'BORE x 1/4 (6.3) DEEP—5 REQ'D.

11.) "LEADER"

3.) 1.00-.015 = .985 (Lower Limit)

8.) "ROUND"

.1.750 (44.45) DIA. 4.)

(1/2 DIA)

2.) .3 (762) DIA.

7.) 3 1/4 - 1 1/2 = 1 3/4 (44.4)

1 1/2 (1/2 DIA)

3 1/4 (82.5)

4.750 (120.65)

7/8 (22.2)

13.) "LEADER"

6.) "RUNOUT"

1 7/8 (47.6)

11/16 (17.4)

11/32 (8.7)

1.188 (30.2)

2.375 (60.4)

3 3/4 (95.2)

DRAWN PBY	1/3/80	CHECKED JCB	1/5/80	INCH MILLIMETER
APPROV. JBC	1/6/80	MATERIAL CAST IRON		

12.) 3rd angle projection (U.S./Canada)

PLATE-END

JAN. ENGINEERING
Box 52
LYNDONVILLE, VT.

DRAWING NO.
A49372-A

DO NOT SCALE DRAWING

TOLERANCE UNLESS OTHERWISE SPEC.
.XX ± .015
.XXX ± .005
.XXXX ± .0002

HEAT TREATMENT

SCALE HALF SIZE

15.) WAS 1 1/16 (26.9) DP. 7/18/80

LET	CHANGE	DATE

CHANGE NOTICE

A WAS 1 1/16 (26.9) DP. 7/18/80

CONVENTIONAL PRACTICES WORKSHEET ANSWERS

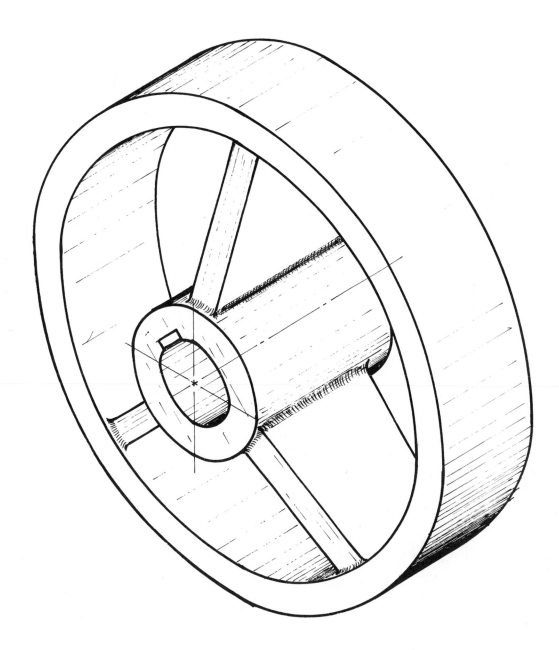

Pictorial view

A549923

CONVENTIONAL PRACTICES WORKSHEET ANSWERS

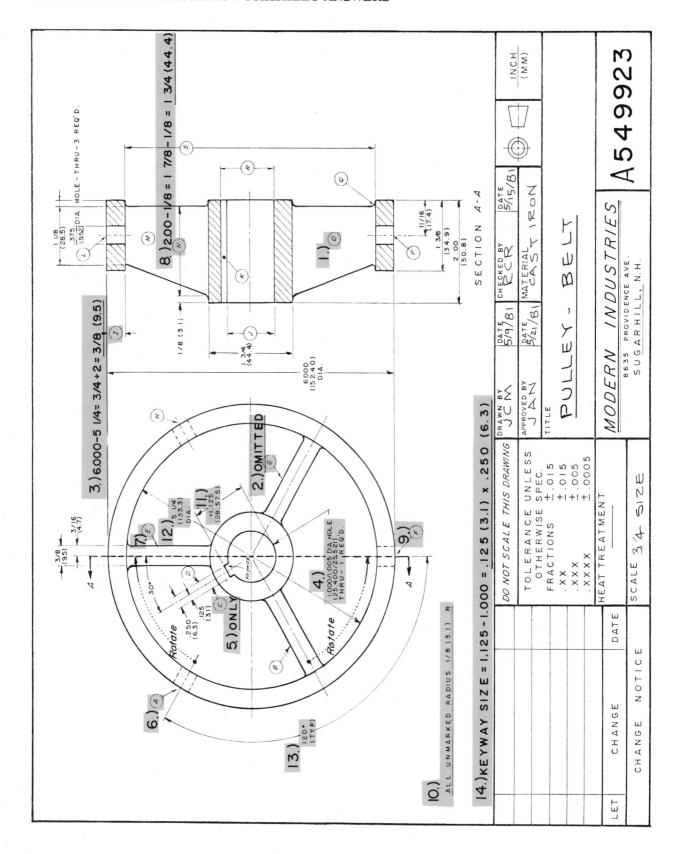

EVALUATION ANSWERS

1. Shafts, bolts, nuts, rods, rivets, keys, pins, spokes, arms, webs, teeth of a gear, screws, bail bearings, roller bearings, etc.
2. No. Only if necessary.
3. No.
4. Each half would be different.
5. 45°
6. Cast iron.
7. Half section.
8. Direction of site.
9. For LONG objects, so you can keep the scale as *large* as possible.
10. It gives the illusion the object is solid all the way around or through.
11. A *broken-out section* view is used to illustrate only one area or one important feature.
12. An *offset section* view is similar to a full section *except* the cutting plane line *bends* in order to illustrate important features.
13. The three reasons rounds and fillets are used are:

    1. Appearance,
    2. Strength, and
    3. To eliminate sharp corners.

14. A *removed section* view is similar to a rotated section.
15. An *assembly section* illustrates various parts of an assembly.
16. Ribs and webs are *not* sectioned so as not to give the illusion the whole object is thick.
17. Holes are sometimes located incorrectly in order to better illustrate the part and be less confusing.
18. Spokes are not sectioned so as to *not* create the illusion the part is solid.

South St. Paul Public Library
106 Third Avenue North
th St. P   MN 55075

AUXILIARY
VIEWS

Objective: To read and interpret various kinds of auxiliary views that are used in industry.

TERMS TO KNOW AND UNDERSTAND

See end of chapter for answers.

Inclined Surfaces_____

_____

_____

_____

Auxiliary View Functions (List Two) _____

_____

_____

_____

Direction Of Sight_____

_____

_____

_____

Kinds Of Auxiliary Views (List Three)_____

_____

_____

_____

_____

AUXILIARY VIEWS

Some objects have one or more surfaces inclined or at an angle. In order to show the true size and shape an auxiliary view is drawn in conjunction with the regular views. Study this regular three-view drawing, notice how one surface is inclined, and yet nowhere is its true size or shape shown.

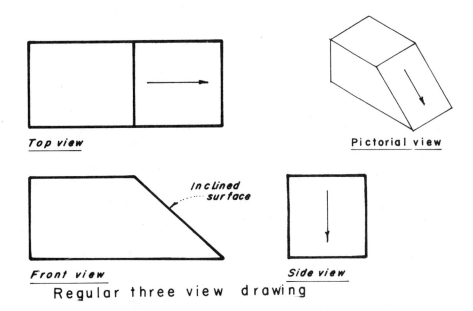

Top view

Pictorial view

Front view

Side view

Regular three view drawing

Here, an auxiliary view is added. Notice the true length of the inclined surface only appears in the auxiliary view.

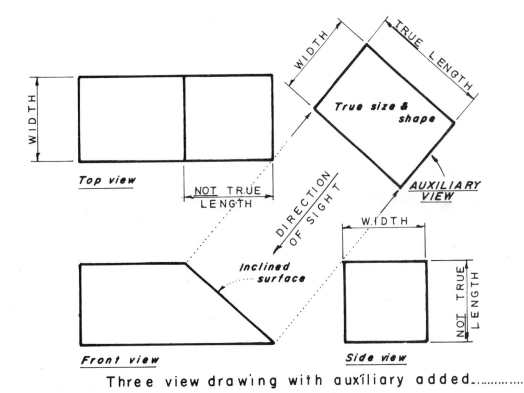

Three view drawing with auxiliary added..............

The auxiliary view is exactly what is seen as indicated by the "direction of sight" arrow. *Only the surface that is on an incline is drawn in the auxiliary view.*

There are three kinds of auxiliary views, front, top, and side view auxiliary. Each is, as its name implies, an auxiliary projected from either the front, top, or side view. An auxiliary view does two things, it shows the true size and shape of an inclined surface. Study each kind of auxiliary view in order to readily be able to identify each.

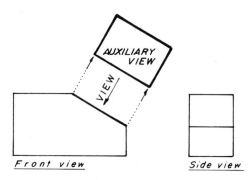

1. FRONT VIEW AUXILIARY

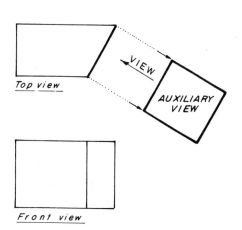

2. TOP VIEW AUXILIARY

An auxilary view does two things, it shows true size and shape of an inclined surface.

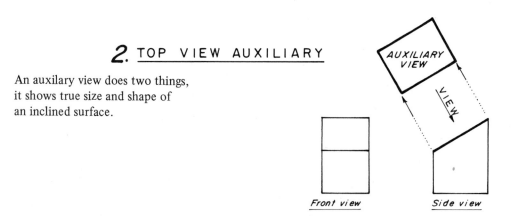

3. SIDE VIEW AUXILIARY

Carefully study the pictorial view of a simple object. Compare it to the three-view drawing below. Notice how surface 'E' in the top view is *not* the true size. Surface 'E' in the side view is *not* the true size either. The only place the true size and shape of surface 'E' is shown, is in the auxiliary view.

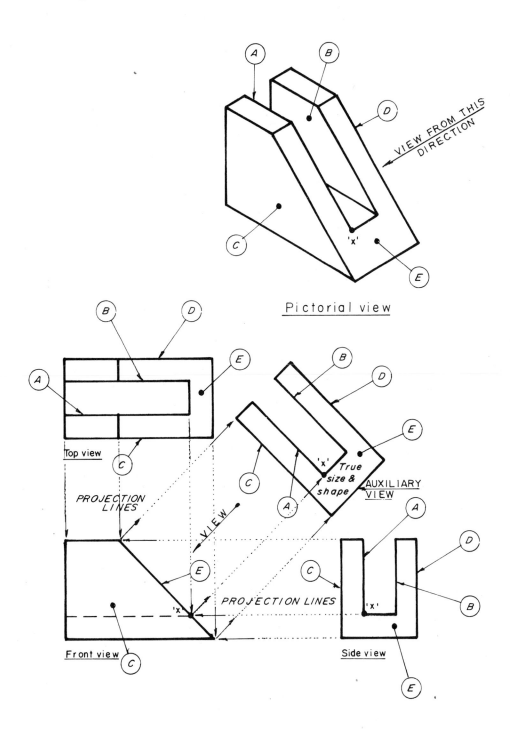

Find surfaces *A, B, C, D,* and *E* in the three regular views and then relate each surface to the *auxiliary view.* Locate point *"X"* in the side view, project it to the *front* view, via projection lines, then up to the auxiliary view.

AUXILIARY VIEWS WORKSHEET I

Using drawing No. A225932 answer the following questions in the space provided. See end of chapter for answers.

1. Surface 'A' in the top view is what surface in the right side view? _____
   _____

2. What is radius 'B'? _____

3. Surface 'C' in the top view is what surface in the auxiliary view? _____
   _____

4. Surface 'E' in the top view is what surface in the auxiliary view? _____
   _____

5. What is dimension 'F'? _____

6. What is radius 'G'? _____

7. Calculate dimension 'H'. _____

8. Calculate dimension 'I'! _____

9. Calculate dimension 'J'. _____

10. Surface 'K' in the front view is what surface in the auxiliary view? _____
    _____

11. What is dimension 'S'? _____

12. What is the *true length* of dimension 'T'? _____
    (As measured along the surface)

13. What is the *true length* of dimension 'U'? _____
    (As measured along the surface)

14. Surface 'V' is the right side view is what surface in the auxiliary view? ____
    _____

15. What is dimension 'Y'? _____

16. What is dimension 'Z'? _____

17. What is the angle between surface 'A' and surface 'C'? _____

18. There are *two* hexagonal surfaces (six-sided) which is probably the most important? Why? _____
    _____

AUXILIARY VIEWS

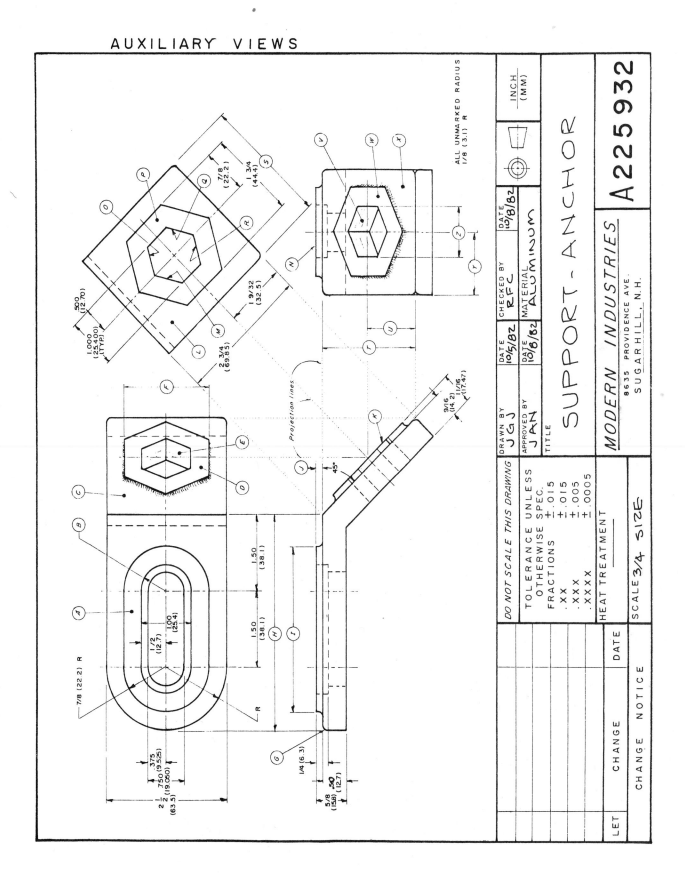

ALL UNMARKED RADIUS
1/8 (3.1) R

INCH
(MM)

SUPPORT-ANCHOR

MODERN INDUSTRIES

A225932

8635 PROVIDENCE AVE.
SUGARHILL, N.H.

| DRAWN BY | JGJ | | DATE | 10/5/82 | CHECKED BY | RFC | DATE | 10/8/82 |
| APPROVED BY | JAN | | DATE | 10/8/82 | MATERIAL | ALUMINUM | | |

TITLE

DO NOT SCALE THIS DRAWING

TOLERANCE UNLESS
OTHERWISE SPEC.
FRACTIONS ±.015
.XX   ±.015
.XXX   ±.005
.XXXX ±.0005

HEAT TREATMENT

SCALE 3/4 SIZE

CHANGE NOTICE

| LET | CHANGE | DATE |

AUXILIARY VIEWS WORKSHEET II

Using drawing A875118 answer the following questions in the space provided. See end of chapter for answers.

1. What is radius 'A'? _____

2. Surface 'B' in the front view is what surface in the right side view? _____
    _____

3. What is radius 'D'? _____

4. What is the *smallest* size dimension 'E' could be and still be within tolerance?
    _____
    _____

5. Surface 'F' in the front view is what surface in the auxiliary view? _____
    _____

6. What is dimension 'G'? _____

7. What is dimension 'H'? _____

8. Surface 'I' in the auxiliary view is what surface in the front view? _____
    _____

9. What is dimension 'K'? _____

10. Surface 'M' in the right side view is what surface in the front view? _____
    _____

11. What is dimension 'O'? _____

12. Calculate radius 'P'. _____

13. What is dimension 'R'? _____

14. What are the limits of the .312 (7.940) dia. holes? _____

15. What was the *length* of the object before it was bent? _____

AUXILIARY VIEWS

DEVELOPED LENGTH = 3.469 (88.210)

1/8 (3.1) R (TYP.)

11/16
(17.5)

120°

1 1/4 (31.7)

9/16
(12.3)

1/4 (6.3)

1 5/16
(33.3)

.50
(12.5)

.312 (7.940) DIA. HOLE
THRU
4 REQ'D.

9/16
(14.3)

2 7/16
(70.3)

R

.25
(6.3)

9/16 (14.3) R

.500
(12.7)

1/2
(12.7)

.828
(21.035)

1.656
(42.070)

2.484
(63.105)

3.50
(88.90)

	DATE 8/5/81	CHECKED BY EPR	DATE 8/10/81	INCH (MM)
DRAWN BY PRR				
APPROVED BY CWM	DATE 8/12/81	MATERIAL STEEL PLATE		

TITLE

PLATE - ADJUSTING

MODERN INDUSTRIES
8635 PROVIDENCE AVE.
SUGARHILL, N.H.

A 875118

DO NOT SCALE THIS DRAWING

TOLERANCE UNLESS
OTHERWISE SPEC.
FRACTIONS ± .015
.XX   ± .015
.XXX   ± .005
.XXXX ± .0005

HEAT TREATMENT

SCALE 3/4 SIZE

LET | CHANGE | DATE

CHANGE NOTICE

AUXILIARY VIEWS WORKSHEET III

Using drawing A675134 answer the following questions in the space provided. See end of chapter for answers.

1. Surface 'C' in the top view is what surface in the front view? _____

2. What is the radius at 'D'? _____

3. Surface 'E' in the top view is what surface in the auxiliary view? _____
   _____

4. Surface 'F' in the top view is what surface in the front view? _____
   _____

5. Surface 'H' in the auxiliary view is what surface in the front view? _____
   _____

6. What is dimension 'J'? _____

7. What is dimension 'K'? _____

8. Surface 'L' in the auxiliary view is what surface in the front view? _____
   _____

9. Surface 'M' in the auxiliary view is what surface in the front view? _____
   _____

10. What is the radius at 'N'? _____

11. Surface 'O' in the front view is what surface in the top view? _____

12. What is point 'R' called? _____

13. Surface 'T' in the front view is what surface in the auxiliary view? _____
    _____

14. What is the actual size of dimension 'U'? _____

15. What is dimension 'V'? _____

16. What dimension 'X'? _____

17. What is dimension 'Y'? _____

18. What is the distance from surface 'A' to surface 'I'? _____

19. What is the distance from surface 'O' to surface 'B'? _____

20. What is the distance from surface 'P' to surface 'H'? _____

AUXILIARY VIEWS

AUXILIARY VIEWS WORKSHEET IV

Using drawing A23194, answer the following questions in the space provided. See end of chapter for answers.

1. Surface 'A' in the front view is what surface in the top view? _____
    _____

2. What is dimension 'C'? _____

3. What is radius 'D'? _____

4. What is the true length of dimension 'F'? _____

5. What is radius 'I'? _____

6. What is dimension 'K'? _____

7. What is dimension 'L'? _____

8. What is the diameter of the pan (M)? _____

9. Surface 'N' in the top view is what surface in the front view? _____

10. What is diameter 'P'? _____

11. What is dimension 'Q'? _____

12. How far apart are the two .375 (9.525) dia. holes? _____

13. What is dimension 'R'? _____

14. Calculate dimension 'S'. _____

15. Surface 'T' in the auxiliary view is what surface in the front view? _____
    _____

16. Surface 'U' in the auxiliary view is what surface in the top view? _____
    _____

17. Surface 'V' in the auxiliary view is what surface in the front view? _____
    _____

18. Surface 'N' is how far from surface 'V'? _____

19. The .375 (9.525) dia. holes are at what angle from the .562 (14.290) dia. hole? _____

20. At what angle is surface 'B' to surface 'T'? _____

.562 (14.290) DIA. HOLE
THRU - 1 REQ'D.

1.25 (31.7) DIA.

1 3/4
(44.4)

7/8
(22.2)

ALL UNMARKED RADIUS
1/8 (3.1) R

.375(9.525) DIA. HOLE
1.00(25.4) DEEP
2 REQ'D.

.750
(19.05)

.50
(12.7)

5/16 (7.9) DIA. HOLE
3/4 (19.0) DEEP
1 REQ'D.

1 5/8
(41.2)

2 3/4
(69.8)

3/8
(9.5)

1/2
(12.7)

45°

1.00
(50.4)

1/2
(12.7)

13/32
(10.3)

30°

1/2
(38.1)

3/8
(9.5)

3/16
(20.6)

R

.312 (7.94)

9/16
(14.2)

1 1/16
(26.9)

2 1/4
(57.1)

.625
(15.88)

R P Q O N M L K J I T S A B C D E F G H U V

DO NOT SCALE THIS DRAWING	DRAWN BY JAN	DATE 7/81	CHECKED BY TWN	DATE 7/81	INCH (MM.)
TOLERANCE UNLESS OTHERWISE SPEC. FRACTIONS ± .015 .XX ± .015 .XXX ± .005 .XXXX ± .0005	APPROVED BY TO B	DATE 8/81	MATERIAL BRASS		
	TITLE ADAPTER - SWIVEL				
HEAT TREATMENT	MODERN INDUSTRIES				A23194
	8635 PROVIDENCE AVE. SUGARHILL, N.H.				
SCALE 3/4 SIZE					

LET	CHANGE	DATE

CHANGE NOTICE

AUXILIARY VIEWS EVALUATION

Using drawing A92115 answer the following questions, in the space provided. See end of chapter for answers.

1. Surface 'A' in the front view is what surface in the top view? _____
_____

2. What is dimension 'B'? _____

3. What size is hole 'C'? _____

4. Surface 'E' in the front view is what surface in the top view? _____
_____

5. Surface 'H' in the front view is what surface in the top view? _____
_____

6. What is the thickness of 'I'? _____

7. Surface 'J' in the left side view is what surface in the front view? _____
_____

8. Surface 'M' in the left side view is what surface in the front view? _____
_____

9. What is radius 'N'? _____

10. Surface 'P' in the auxiliary view is what surface in the left side view? _____
_____

11. What kind of line is the line indicated by 'O'? _____

12. What is radius 'Q'? _____

13. Calculate dimension 'S'. _____

14. Surface 'U' in the top view is what surface in the front view? _____

15. What is radius 'V'? _____

16. What is the actual length of the leg at 'W'? _____

17. What is the *maximum* size dimension 'X' can be and still be within tolerance?
_____

18. Surface 'Y' in the auxiliary view is what surface in the front view? _____
_____

19. Surface 'R' is what angle from surface 'T'? _____

20. How far from surface 'T' is the center location of the .812 (20.640) dia. hole? _____

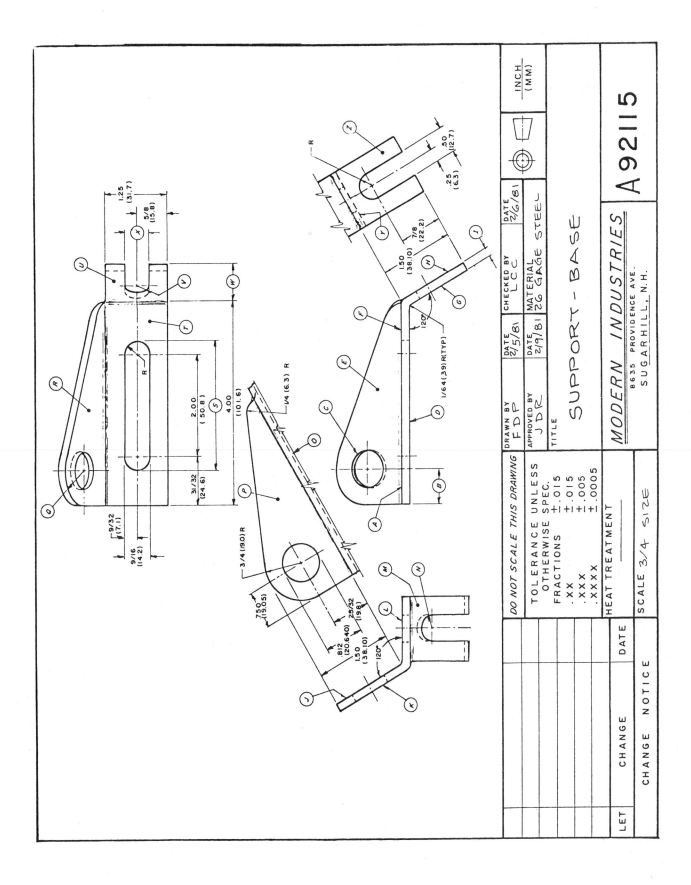

ANSWERS TO TERMS TO KNOW AND UNDERSTAND

Inclined Surfaces—A surface on an *angle*, not 90° to the usual surfaces.

Auxiliary View Functions (2)—True *size* and true *shape*.

Direction Of Sight—The angle the surface is viewed from (90° from the surface).

Kinds Of Auxiliary Views—Front auxiliary view, side auxiliary view, and top auxiliary view.

AUXILIARY VIEWS WORKSHEET I ANSWERS

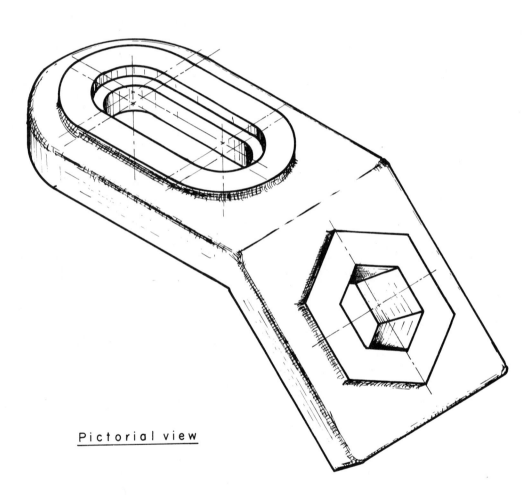

Pictorial view

A225932

AUXILIARY VIEWS WORKSHEET I ANSWERS

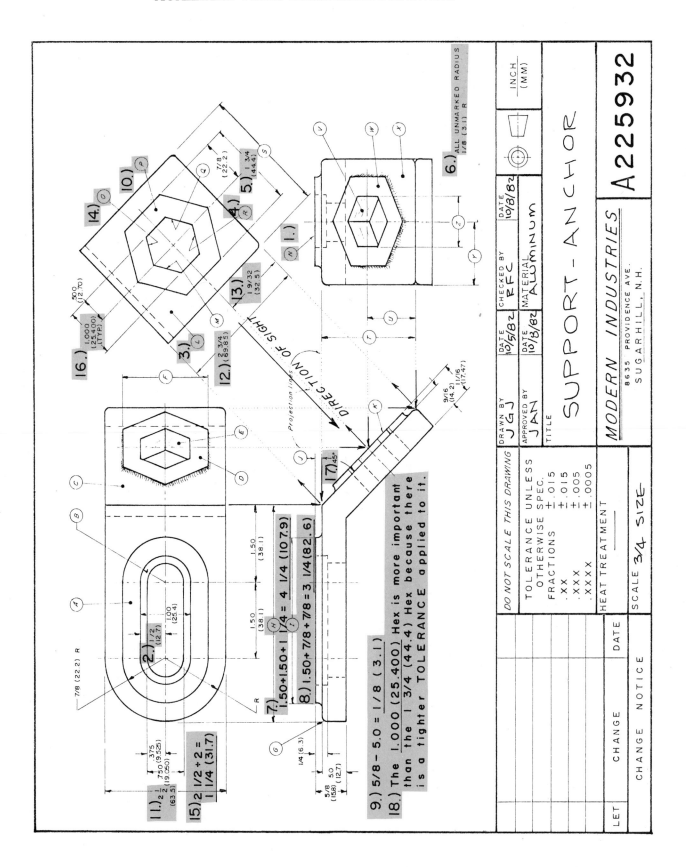

6.) ALL UNMARKED RADIUS 1/8 (3.1) R

5.) 1 3/4 (44.4)

7/8 (22.2)

.500 (12.70)

1.000 (25.400) (TYP.)

3.) 9/32 (32.5)

12.) 2 3/4 (69.85)

F

DIRECTION OF SIGHT

Projection lines

17.) 45°

1.50 (38.1)

1.50 (38.1)

7/8 (22.2) R

.375 (9.525)

.750 (19.050)

1.00 (25.4)

2.) 1/2 (12.7)

11.) 2 1/2 (63.5)

15.) 2 1/2 + 2 = 1 1/4 (31.7)

7.) 1.50+1.50+1 1/4 = 4 1/4 (107.9)

8.) 1.50 + 7/8 + 7/8 = 3 1/4 (82.6)

9.) 5/8 - 5.0 = 1/8 (3.1)

18.) The 1.000 (25.400) Hex is more important than the 1 3/4 (44.4) Hex because there is a tighter TOLERANCE applied to it.

1/4 (6.3)

5/8 (15.8)

5.0 (12.7)

9/16 (14.2)

11/16 (17.47)

DO NOT SCALE THIS DRAWING

TOLERANCE UNLESS OTHERWISE SPEC.
FRACTIONS ± .015
.XX ± .015
.XXX ± .005
.XXXX ± .0005

HEAT TREATMENT

DRAWN BY JGJ

DATE 10/5/82

CHECKED BY RFC

DATE 10/8/82

APPROVED BY JAN

DATE 10/13/82

MATERIAL ALUMINUM

INCH (MM)

A225932

TITLE

SUPPORT - ANCHOR

MODERN INDUSTRIES

8635 PROVIDENCE AVE.
SUGAR HILL, N.H.

SCALE 3/4 SIZE

LET | CHANGE | DATE

CHANGE NOTICE

AUXILIARY VIEWS WORKSHEET II ANSWERS

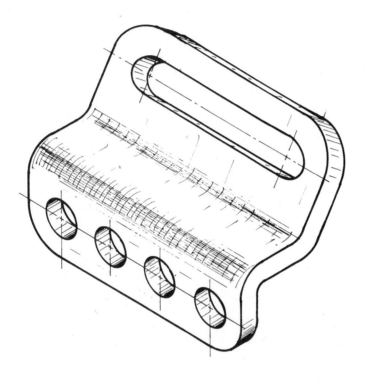

Pictorial view

A 875118

AUXILIARY VIEWS WORKSHEET II ANSWERS

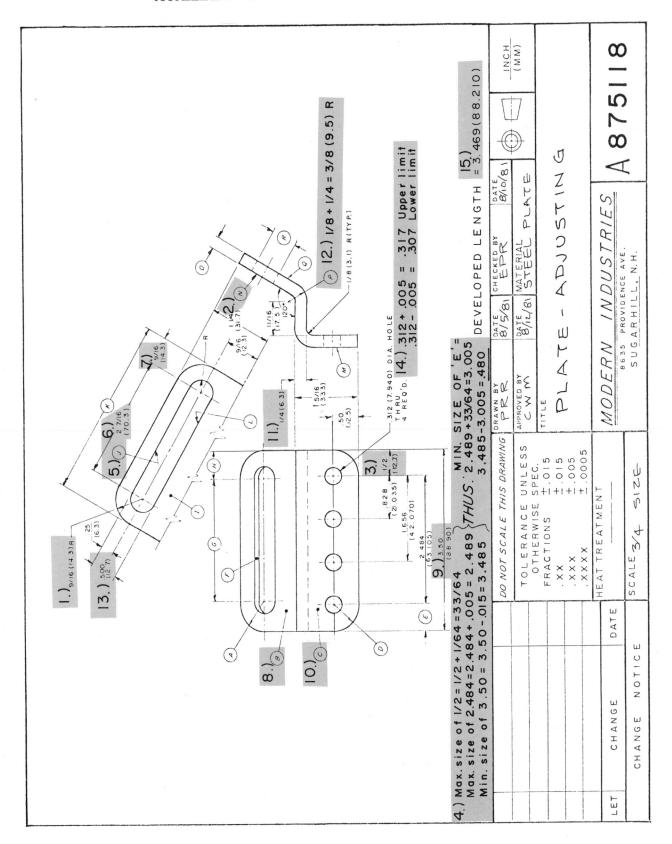

1.) 9/16 (14.3) R

13.) .500 (12.7)

8.)

10.)

5.)

6.) 2 7/16 (70.3)

7.) 9/16 (14.3)

11.) 1/4 (6.3)

2.) 1 1/4 (31.7)

12.) 1/8 + 1/4 = 3/8 (9.5) R
1/8 (3.1) R (TYP.)

3.) 1/2 (12.7)

9.) 3.50 (88.90)

.312 (7.940) DIA. HOLE
THRU
4 REQ'D.

14.) .312 + .005 = .317 Upper limit
.312 − .005 = .307 Lower limit

15.) 3.469 (88.210)

DEVELOPED LENGTH =

4.) Max. size of 1/2 = 1/2 + 1/64 = 33/64
   Max. size of 2.489 = 2.489 + 33/64 = 3.005
   Min. size of 2.484 = 2.484 + .005 = 2.489 ⎱ THUS: MIN. SIZE OF 'E' =
   Max. size of 2.489 = 2.489 + .005 = 3.005
   3.485 − 3.005 = .480
   Min. size of 3.50 = 3.50 − .015 = 3.485

DO NOT SCALE THIS DRAWING

TOLERANCE UNLESS
OTHERWISE SPEC.
FRACTIONS ±.015
   .XX   ±.015
   .XXX   ±.005
   .XXXX   ±.0005

HEAT TREATMENT _____

DRAWN BY	DATE
PRR	8/5/81

CHECKED BY	DATE
EPR	8/10/81

APPROVED BY	MATERIAL
C W M	STEEL PLATE
	DATE 8/12/81

TITLE

PLATE − ADJUSTING

MODERN INDUSTRIES
8635 PROVIDENCE AVE.
SUGARHILL, N.H.

A 875118

INCH
(MM)

SCALE 3/4 SIZE

LET	CHANGE	DATE

CHANGE NOTICE

11/16 (17.5)
120°

9/16 (14.3)

1 5/16 (33.3)

.50 (12.5)

.828 (21.035)

1.656 (42.070)

2.484 (63.105)

.25 (6.3)

1 5/16 (33.3)

AUXILIARY VIEWS WORKSHEET III ANSWERS

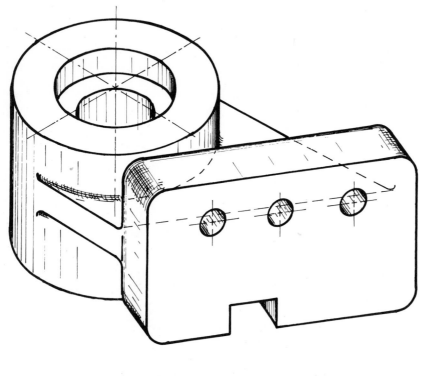

Pictorial view

A 675134

AUXILIARY VIEWS III WORKSHEET ANSWERS

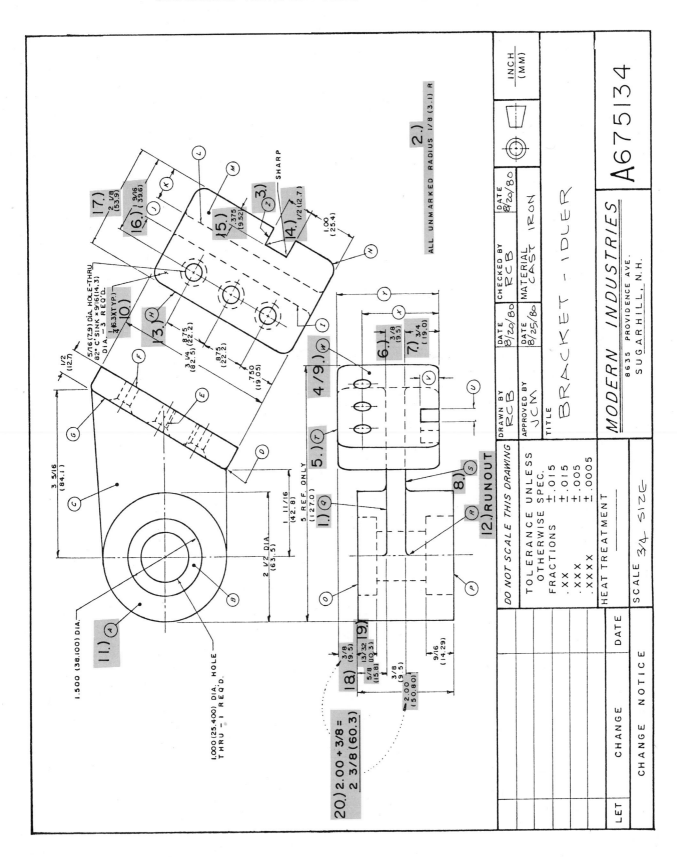

AUXILIARY VIEWS WORKSHEET IV ANSWERS

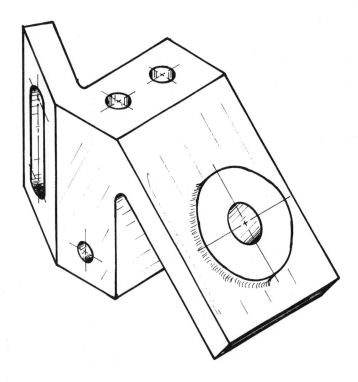

Pictorial view

A23194

AUXILIARY VIEWS WORKSHEET IV ANSWERS

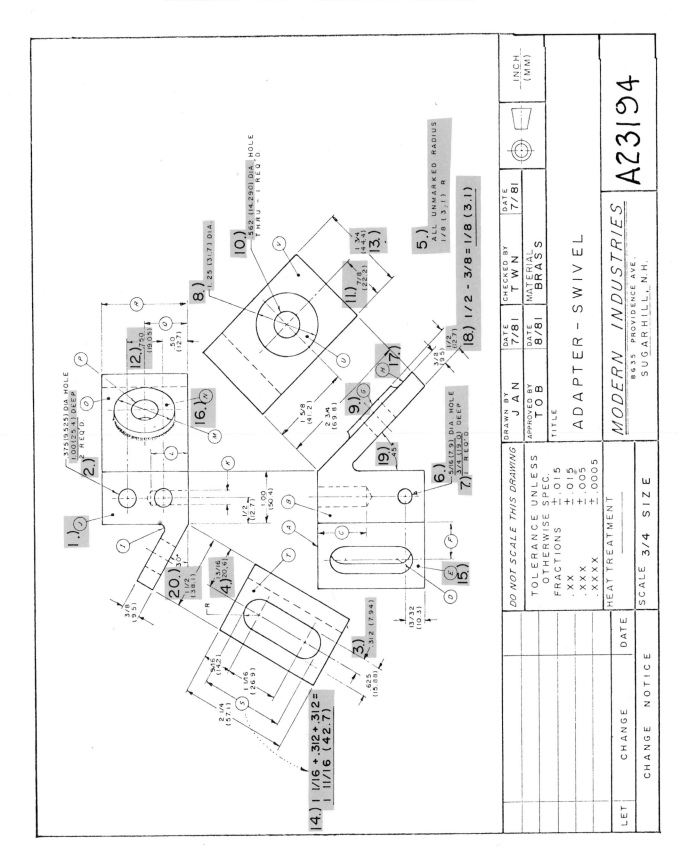

.562 (14.290) DIA. HOLE
THRU - 1 REQ'D.

1.25 (31.7) DIA.

1 3/4
(44.4)

ALL UNMARKED RADIUS
1/8 (3.1) R

7/8
(22.2)

.750
(19.05)

.50
(12.7)

.375 (9.525) DIA HOLE
1.00 (25.4) DEEP
2 REQ'D.

1/2 - 3/8 = 1/8 (3.1)

1/2
(12.7)

3/8
(9.5)

1 5/8
(41.2)

2 3/4
(69.8)

5/16 (7.9) DIA HOLE
3/4 (19.0) DEEP
1 REQ'D.

45°

1.00
(50.4)

1/2
(12.7)

3/8
(9.5)

1/2
(38.1)

30°

3/16
(20.6)

R

13/32
(10.3)

.312 (7.94)

9/16
(14.2)

1 1/16
(26.9)

.625
(15.88)

2 1/4
(57.1)

1 1/16 + .312 + .312 =
1 11/16 (42.7)

DO NOT SCALE THIS DRAWING

DRAWN BY JAN	DATE 7/81	CHECKED BY TWN	DATE 7/81
APPROVED BY TOB	DATE 8/81	MATERIAL BRASS	

TOLERANCE UNLESS
OTHERWISE SPEC.
FRACTIONS ± .015
.XX ± .015
.XXX ± .005
.XXXX ± .0005

HEAT TREATMENT

SCALE 3/4 SIZE

TITLE

ADAPTER – SWIVEL

MODERN INDUSTRIES

8635 PROVIDENCE AVE.
SUGARHILL, N.H.

A23194

INCH
(MM)

LET	CHANGE	DATE

CHANGE NOTICE

AUXILIARY VIEWS EVALUATION ANSWERS

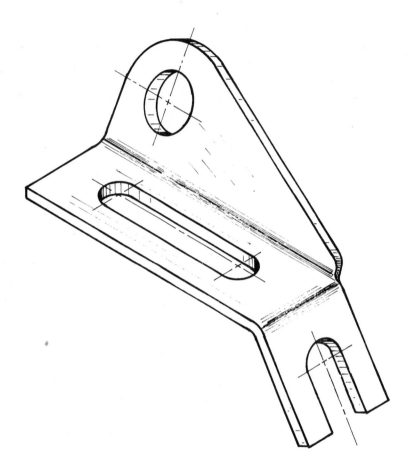

Pictorial view

A 92115

AUXILIARY VIEWS EVALUATION ANSWERS

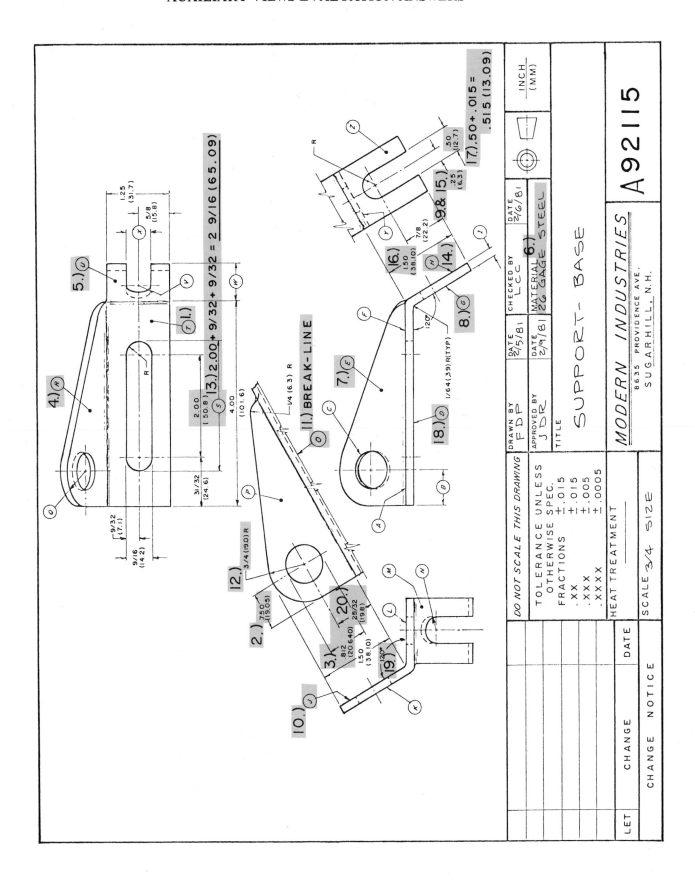

UNIT 7

MISCELLANEOUS INFORMATION

Objective: To interpret drawings which depict various drafting "practices" used today in industry.

TERMS TO KNOW AND UNDERSTAND

See end of chapter for answers.

Tap _____

Die _____

Permanent Fasteners _____

Temporary Fasteners _____

Purposes Of Threads _____

Unified National Thread _____

Square, Acme, Buttress, And Worm Threads _____

Major/Minor Dia. Of Threads _____

Pitch _____

Thread Call-off _____

T.P.I. _____

Keyway _____

Kinds Of Keys _____

"Ref" Dimension _____

Chamfer _____

Knurling _____

Undercut (relief) _____

Machine Screws _____

Cap Screws _____

Finish Marks _____

M.M.C. _____

Allowance _____

Clearance _____

Nominal Size (nom.) _____

Types Of Welds _____

Types Of Joints _____

Welding Symbol _____

Alloy _____

Metallurgy _____

Strength _____

Plasticity _____

Ductility _____

Malleability _____

Elasticity _____

Brittleness _____

Toughness _____

Fatigue Limit _____

Aging _____

Quenching _____

Tempering _____

Annealing _____

Case Hardening _____

Extrusion _____

Drilling _____

Turning _____

Planing _____

Grinding _____

Milling _____

Lathe _____

Turret _____

Broach _____

Casting _____

Split Pattern _____

Core Support _____

Molding Board _____

Drag _____

Cope _____

Flask _____

Sprue _____

Riser _____

Shrinkage Allowance _____

Draft _____

Bosses And Pads _____

Lugs _____

FASTENERS

Anything that is to be assembled usually is held together by a fastener. There are two major classifications of fasteners. The first is designed to *permanently* join parts together. The second is designed to permit assembly and disassembly.

1. *Permanent Fasteners*
   - Welding
   - Brazing
   - Stapling
   - Nailing
   - Gluing
   - Riveting

2. *Temporary Fasteners* (For assembly or disassembly)
   - Screws
   - Bolts
   - Keys
   - Pins

Threads are used for four purposes:

1. Fastening
2. Adjusting
3. Transmitting power
4. Measuring

TAP AND DIE

There are various methods used to produce inside and outside threads. The simplest method is the use of a tap (internal threads) or a die (external threads). Taps and dies come in all standard sizes (dia.) and all standard T.P.I. sizes.

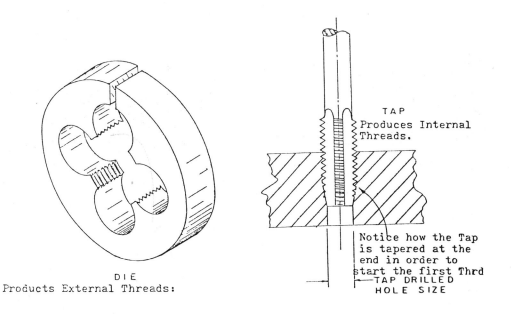

DIE
Products External Threads:

TAP
Produces Internal Threads.

Notice how the Tap is tapered at the end in order to start the first Thrd
TAP DRILLED HOLE SIZE

THREAD PROFILE

The shape or profile of a thread depends upon its function. The most used shape is the unified and american national. Square, acme, worm, and buttress threads are used primarily to transmit power (buttress in only one direction). (Note: "Pitch" determines all other dimension).

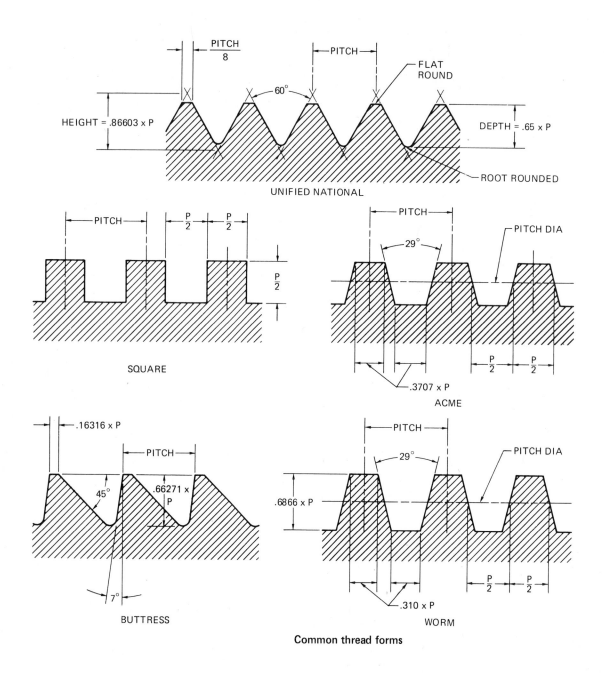

Common thread forms

THREAD TERMS

There are various terms associated with threads. It is important that the crafts-person understands these terms. Each important thread term has been illustrated below and on the next few pages.

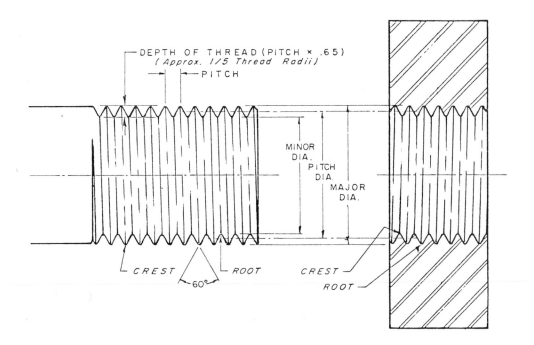

Thread Terms:

Minor Dia.—(External Threads): from root to root. (Internal Threads: from crest to crest.

Major Dia.—(External Threads): from crest to crest. (Internal Threads): from root to root.

Crest—(External Threads): high point on threads. (Internal Threads): inside point on threads.

Root—(External Threads): low point on threads. (Internal Threads): high point on threads.

Pitch Dia.—A diameter located halfway between the major dia. and the minor dia. Thread angle at 60° (Unified National).

Pitch—Distance from one thread to the exact same location on the next thread.

Depth of Thread = pitch x 0.65.

THREAD REPRESENTATION

There are two kinds of threads, *external,* such as on a bolt and *internal,* as on a nut.

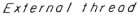

External thread

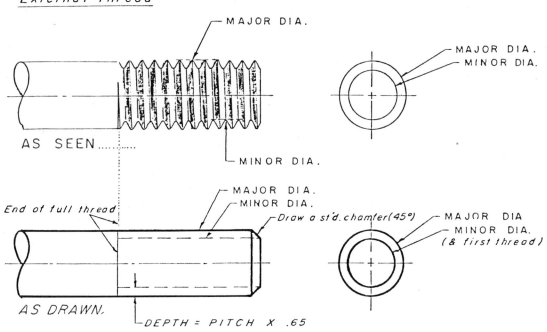

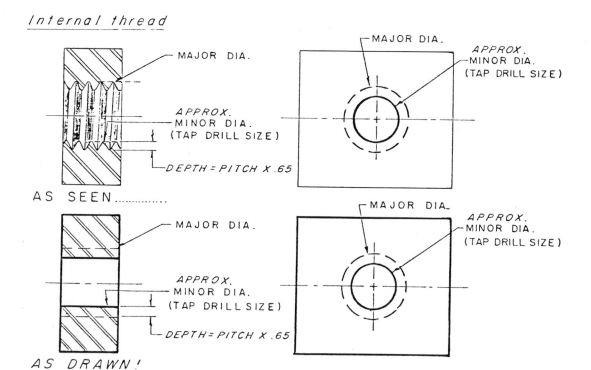

TAPPED THREADED BLIND HOLE

A blind hole, is a hole that does not go through an object.

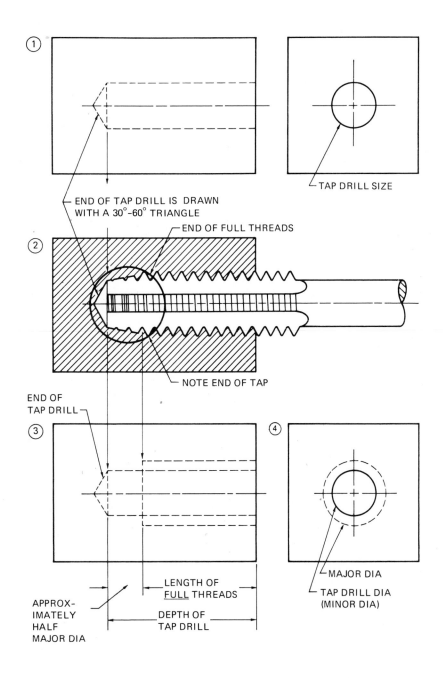

THREAD CALL-OFF

Companies use various methods to call-off a thread or fastener size. Noted below is a recommended method call-off:

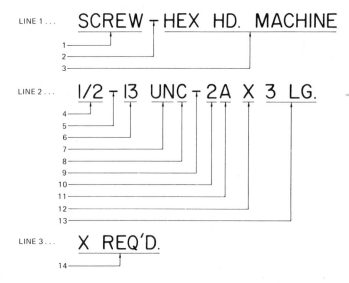

Line 1

1. Call-off what it is; i.e., screw, nut, washer, pin, etc. (think of the first item as a noun).
2. The dash indicates where you would start to say what the call-off is, i.e. "Hex Hd. Machine Screw." (Think of everything that follows the dash as an adjective—describing the noun).
3. This part of the call describes the item in full. (Note: This method of call-offs is used in many areas of engineering).

Line 2

4. Nominal size (in inch systems, fractions are used).
5. The dash simply separates.
6. Threads per inch (T.P.I.).
7. U.N. = Unified National *Series.*
8. C = Coarse (F = fine or E.F. = extra fine).
9. The dash simply separates.
10. 2 = indicates class of *fit* (average); 1 = loose fit, 3 = tight fit.
11. A = indicates external thread (B = internal thread).
12. X = used to separate.
13. Last call-off indicates length (if required).

Line 3

14. Indicates how many required.

Examples:

Nut – Hex Hd.	Washer – Lock	Pin – Cotter
5/8 – 18 UNF	3/8 size	1/16 size x 1" Lg.
2 Required	1 Required	10 Required

THREADS OF BOTH SYSTEMS

At this time the U.S.A., Canada, and the United Kingdom use the inch series of Unified Thread Form. (U.N.C./U.N.F.). Using this system (U.N.C./U.N.F.) designates the diameter and *threads* per inch, along with the suffix indicating thread series (coarse or fine) Example:

1/4 – 20 UNC

METRIC THREAD CALL-OFF

Threads in metrics use a similar approach except for a prefix of 'M'. The diameter and *pitch* are used to designate the series of fine thread. Example: M16 x 1.5. The 'M' and the diameter only are used if the series is coarse thread, for example: M16.

This could be confusing. In the inch system, fine threads have more T.P.I. than coarse threads, thus a *higher* number. Using the metric system, *pitch* is used in place of T.P.I., thus a fine thread would have a *lower* number. The metric system uses coarse thread more than fine thread. For example:

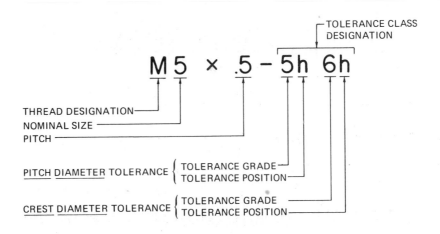

5 MM diameter/.5 Pitch/ with a pitch diameter tolerance of 5 and allowance "h"/a crest diameter tolerance of 6 and allowance "h". This completely describes the threads required.

Note: "h" or "g" (lower case) indicate *external threads.*
    "H" or "G" (upper case) indicate *internal threads.*

HOW TO MEASURE THREADS

Threads per inch (T.P.I.) can easily be measured by the following methods:

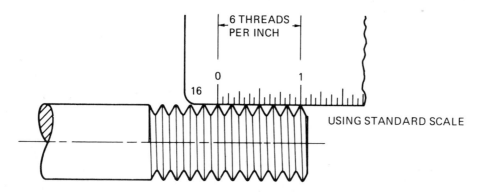

If only *part* of an inch is threaded, count crests in half an inch and double the number to find total T.P.I.

In order to measure T.P.I., use either a standard scale (line up "0" on one crest and count how many crests are in an inch), or to be 100% exact, use a screw pitch gauge as shown.

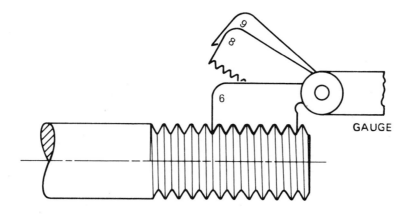

RIGHT HAND AND LEFT HAND THREAD

Threads can be either right hand or left hand. In order to distinguish between a R.H. or L.H. use this simple trick.

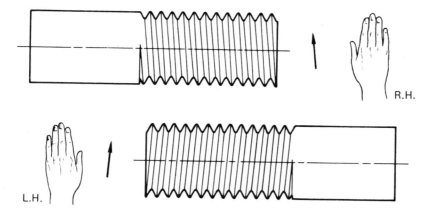

A right hand thread winding will tend to lean to the left as shown. Simply place your hands next to the thread and note how your thumb is naturally leaning.

If the thread leans to the left as in "A"/your Right Hand thumb leans the *same* way, thus it is a R.H. thread. If the thread leans to the right as in "B", your Left Hand thumb leans to the right, thus it is a L.H. thread. *Note:* R.H. threads are not usually indicated, L.H. threads must always be called-off.

USE OF A STANDARDS CHART

Many times in interpreting drawings the craftsperson must read and understand standards charts. On page 198 is a standards chart of *Unified National Threads* which is used in the United States, Canada, and the United Kingdom. (The metric system has been added for reference only).

To use the chart with 1/4 – 20 – UNC threads as an example:

1. Find the nominal size (dia.) of the threads in the left-hand column.
2. Choose either UN*C* (coarse) or UN*F* (fine) thread.
3. Read across at this level for any information you need.

Example: A, 1/4 – 20 UN*C* thread, has a *major* dia. of .250 (6.350)/*minor* dia. of .1850 (4.699) the tap drill is a #7 (.2010).

"Pitch" is the distance from a point on one thread to the exact same point on the next thread. (The pitch for a 1/4 – 20 UNC thread is .050 = 1.000 inch ÷ 20 T.P.I.

STANDARDS CHART

South St. Paul Public Library
106 Third Avenue North
South St. Paul, MN 55075

STANDARDS

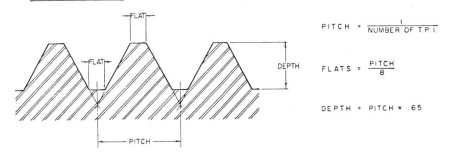

$$\text{PITCH} = \frac{1}{\text{NUMBER OF T.P.I.}}$$

$$\text{FLATS} = \frac{\text{PITCH}}{8}$$

$$\text{DEPTH} = \text{PITCH} \times .65$$

DIMENSION AND SIZE CHART FOR THREADS

Nominal Size		Diameter (Major)		Diameter (Minor)		Tap Drill (For 75% Th'd.)			Threads Per Inch		Pitch (MM)		T.P.I. (Approx.)	
Inch	M.M.	Inch	M.M.	Inch	M.M.	Drill	Inch	M.M.	UNC	UNF	Coarse	Fine	Coarse	Fine
—	M1.4	.055	1.397	—	—	—	—	—	—	—	.3	.2	85	127
0	—	.060	1.524	.0438	1.092	3/64	.0469	1.168	—	80	—	—	—	—
—	M1.6	.063	1.600	—	—	—	—	—	—	—	.35	.2	74	127
1	—	.073	1.854	.0527	1.320	53	.0595	1.499	64	—	—	—	—	—
1	—	.073	1.854	.0550	1.397	53	.0595	1.499	—	72	—	—	—	—
—	M.2	.079	2.006	—	—	—	—	—	—	—	.4	.25	64	101
2	—	.086	2.184	.0628	1.587	50	.0700	1.778	56	—	—	—	—	—
2	—	.086	2.184	.0657	1.651	50	.0700	1.778	—	64	—	—	—	—
—	M2.5	.098	2.489	—	—	—	—	—	—	—	.45	.35	56	74
3	—	.099	2.515	.0719	1.828	47	.0785	1.981	48	—	—	—	—	—
3	—	.099	2.515	.0758	1.905	46	.0810	2.057	—	58	—	—	—	—
4	—	.112	2.845	.0795	2.006	43	.0890	2.261	40	—	—	—	—	—
4	—	.112	2.845	.0849	2.134	42	.0935	2.380	—	48	—	—	—	—
—	M3	.118	2.997	—	—	—	—	—	—	—	.5	.35	51	74
5	—	.125	3.175	.0925	2.336	38	.1015	2.565	40	—	—	—	—	—
5	—	.125	3.175	.9055	2.413	37	.1040	2.641	—	44	—	—	—	—
6	—	.138	3.505	.0975	2.464	36	.1065	2.692	32	—	—	—	—	—
6	—	.138	3.505	.1055	2.667	33	.1130	2.870	—	40	—	—	—	—
—	M4	.157	3.988	—	—	—	—	—	—	—	.7	.35	36	51
8	—	.164	4.166	.1234	3.124	29	.1360	3.454	32	—	—	—	—	—
8	—	.164	4.166	.1279	3.225	29	.1360	3.454	—	36	—	—	—	—
10	—	.190	4.826	.1359	3.429	26	.1470	3.733	24	—	—	—	—	—
10	—	.190	4.826	.1494	3.785	21	.1590	4.038	—	32	—	—	—	—
—	M5	.196	4.978	—	—	—	—	—	—	—	.8	.5	32	51
12	—	.216	5.486	.1619	4.089	16	.1770	4.496	24	—	—	—	—	—
12	—	.216	5.486	.1696	4.293	15	.1800	4.572	—	28	—	—	—	—
—	M6	.236	5.994	—	—	—	—	—	—	—	1.0	.75	25	34
1/4	—	.250	6.350	.1850	4.699	7	.2010	5.105	20	—	—	—	—	—
1/4	—	.250	6.350	.2036	5.156	3	.2130	5.410	—	28	—	—	—	—
5/16	—	.312	7.938	.2403	6.096	F	.2570	6.527	18	—	—	—	—	—
5/16	—	.312	7.938	.2584	6.553	I	.2720	6.908	—	24	—	—	—	—
—	M8	.315	8.001	—	—	—	—	—	—	—	1.25	1.0	20	25
3/8	—	.375	9.525	.2938	7.442	5/16	.3125	7.937	16	—	—	—	—	—
3/8	—	.375	9.525	.3209	8.153	Q	.3320	8.432	—	24	—	—	—	—
—	M10	.393	9.982	—	—	—	—	—	—	—	1.5	1.25	17	20
7/16	—	.437	11.113	.3447	8.738	U	.3680	9.347	14	—	—	—	—	—
7/16	—	.437	11.113	.3726	9.448	25/64	.3906	9.921	—	20	—	—	—	—
—	M12	.471	11.963	—	—	—	—	—	—	—	1.75	1.25	14.5	20
1/2	—	.500	12.700	.4001	10.162	27/64	.4219	10.715	13	—	—	—	—	—
1/2	—	.500	12.700	.4351	11.049	29/64	.4531	11.509	—	20	—	—	—	—
—	M14	.551	13.995	—	—	—	—	—	—	—	2	1.5	12.5	17
9/16	—	.562	14.288	.4542	11.531	31/64	.4844	12.3031	12	—	—	—	—	—
9/16	—	.562	14.288	.4903	12.446	33/64	.5156	13.096	—	18	—	—	—	—
5/8	—	.625	15.875	.5069	12.852	17/32	.5312	13.493	11	—	—	—	—	—
5/8	—	.625	15.875	.5528	14.020	37/64	.5781	14.684	—	18	—	—	—	—
—	M16	.630	16.002	—	—	—	—	—	—	—	2	1.5	12.5	17
—	M18	.709	18.008	—	—	—	—	—	—	—	2.5	1.5	10	17
3/4	—	.750	19.050	.6201	15.748	21/32	.6562	16.668	10	—	—	—	—	—
3/4	—	.750	19.050	.6688	16.967	11/16	.6875	17.462	—	16	—	—	—	—
—	M20	.787	19.990	—	—	—	—	—	—	—	2.5	1.5	10	17
—	M22	.866	21.996	—	—	—	—	—	—	—	2.5	1.5	10	17
7/8	—	.875	22.225	.7307	18.542	49/64	.7656	19.446	9	—	—	—	—	—
7/8	—	.875	22.225	.7822	19.863	13/16	.8125	20.637	—	14	—	—	—	—
—	M24	.945	24.003	—	—	—	—	—	—	—	3	2	8.5	12.5
1	—	1.000	25.400	.8376	21.2598	7/8	.8750	22.225	8	—	—	—	—	—
1	—	1.000	25.400	.8917	22.632	59/64	.9219	23.415	—	12	—	—	—	—
—	M27	1.063	27.000	—	—	—	—	—	—	—	3	2	8.5	12.5

THREADS WORKSHEET

See end of chapter for answers.

1. How are internal threads usually produced? _____
_____

2. List four kinds of temporary fasteners? _____
_____
_____

3. What kinds of thread are used to transmit power? _____
_____

4. List six kinds of permanent fasteners. _____
_____
_____

5. What is the standard angle of the threads of a unified national form thread?
_____
_____

6. How would a 3/8 inch coarse external thread be called-off? (average class of fit). _____
_____

7. How many T.P.I. are there in a #12 *fine* thread? _____

8. What size tap drill would be used on a 9/16-18 UNF thread? _____
_____

9. What is the *minor dia.* of a 1/2-20 UNF thread? _____
_____

10. What is the major diameter in millimeters of a 7/8-14 UNF thread? _____
_____
_____

11. How many *full turns* will it take for a coarse, 1/2 inch dia. thread to travel 1 inch? _____

12. What is the *pitch* of a 3/4 inch coarse thread? _____
_____

KEYWAYS AND KEYSEATS

Keys are used to prevent the rotation of wheels, gears, etc. on their shafts. The key is positioned in a groove on the shaft called a *keyseat*. The key extends above the shaft and fits into the *keyway* cut in the wheel, gear, etc.

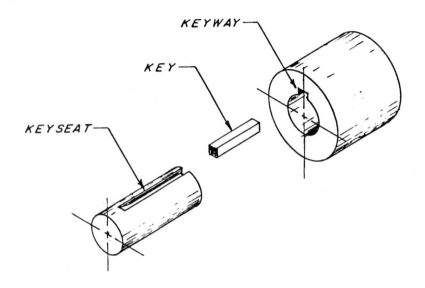

Dimensioning

This is how the *keyseat* and *keyway* should be dimensioned.

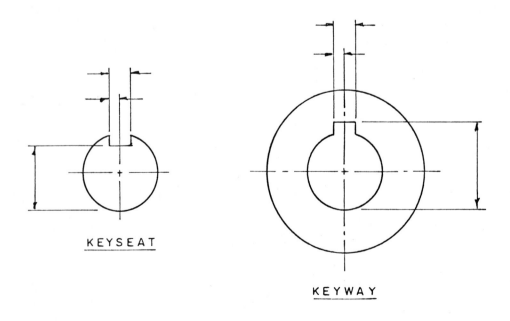

KEYSEAT

KEYWAY

OTHER TYPES OF KEYS

The simplest is the *square* key, illustrated above. Other kinds of keys are the woodruff key, bib-head key and the Pratt & Whitney key. Each kind of key has its own keyseat.

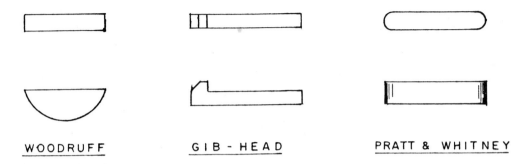

WOODRUFF   GIB - HEAD   PRATT & WHITNEY

REFERENCE DIMENSIONING

A 'reference' dimension is a dimension added for *reference only*. This dimension should *not* be used to manufacture the part.

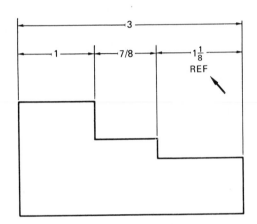

Reference dimensioning

CHAMFERS

Chamfers are usually used at the end of rounded parts. There are two kinds of chamfers, 45° and *other* than 45°.

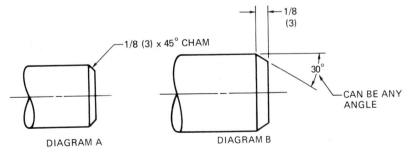

Chamfer dimensioning

KNURLING

Knurling is the process of rolling depressions of either a straight or diamond design into a cylindrical surface.

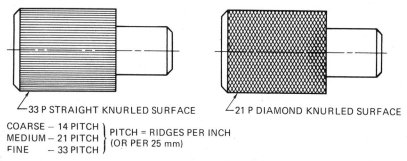

33 P STRAIGHT KNURLED SURFACE   21 P DIAMOND KNURLED SURFACE

COARSE – 14 PITCH
MEDIUM – 21 PITCH   PITCH = RIDGES PER INCH
FINE   – 33 PITCH   (OR PER 25 mm)

Dimensioning knurled surfaces

The more grooves per inch, the finer the knurl. Knurls are used for grip, for a better pressed fit between parts, and for appearance.

Coarse   14
Medium 21   Pitch = Ridges per inch (25.4)
Fine   33

THREAD UNDERCUT (OR RELIEF)

A simple spacer with threads would look like this:

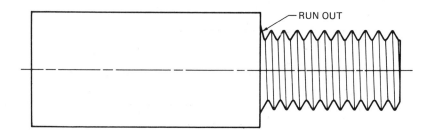

If an arm were attached (and held in place with a nut) there probably would be a space between the shoulder and the arm as shown.

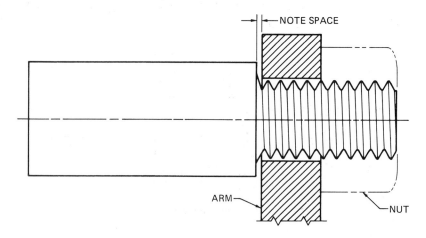

In order to eliminate this possibility, add an undercut or thread relief.

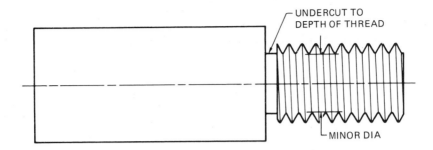

Now, the arm will fit against the shoulder tightly.

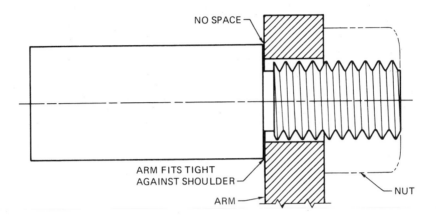

The thread relief is usually designed to the depth of threads, and called off as: *".XX wide X.XX deep thread relief."*

SCREW MACHINE (from 0 to 12 size)

There are four standard types of heads under the classification of *machine screws:* flat, round, oval, and fillister. The machine screw is the smallest screw and starts at #0 size through to a #12 size.

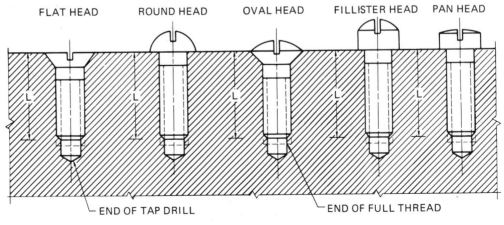

Machine screws

CAP SCREWS (from 1/4 to 1 1/4 size)

There are five standard types of heads under the classification of *cap screws:* flat, round, fillister, hex, and socket head. The cap screw is larger than the machine screw and ranges from 1/4 inch through 1 1/4 inch in size.

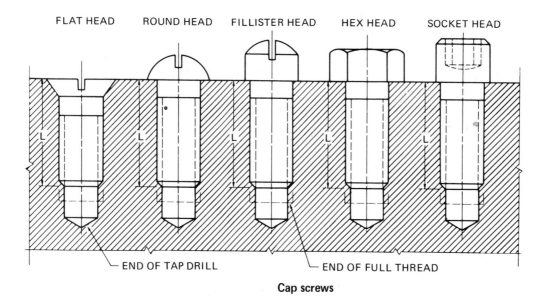

Cap screws

FINISH MARKS

A finished surface is a surface that has extra machining done to it to either make it smooth or to make it flat (or both), usually for mating parts. Throughout the years a finish mark went from:

A finish mark should be made to approximately the size and shape illustrated and put in freehand. A number which equals the roughness you want is neatly printed, 1/8 inch high.

A microinch = .000001 of an inch and simply means:

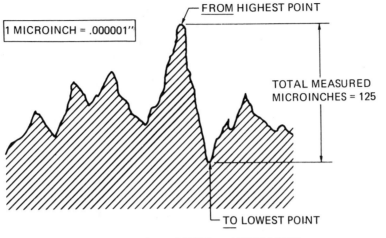

Surface irregularity

ROUGHNESS		KIND OF SURFACE	USAGE
μm	μin.		
12.5	500	Rough	Used where vibration or stress concentration are not critical and close tolerances are not required.
6.3	250	Medium	For general use where stress requirements and appearance are of minimal importance.
3.2	125	Average smooth	For mating surfaces of parts held together by bolts and rivets with no motion between them.
1.6	63	Smoother than average finish	For close fits or stressed parts except rotating shafts, axles, and parts subject to vibrations.
0.8	32	Fine finish	Used for such applications as bearings.
0.4	16	Very fine finish	Used where smoothness is of primary importance such as high-speed shaft bearings.
0.2	8	Extremely fine finish	Use for such parts as surfaces of cylinders (engines).
0.1	4	Super fine finish	Used on areas where surfaces slide and lubrication is not dependable.

**Roughness chart for common finished surface μm = micrometre
μin. = microinch**

Below are some guidelines for finish marks (in microinches):

- Sandcasting—1000 to 300
- Forgings—500 to 100
- Die castings—150 to 16
- Extruded shapes—150 to 10
- Drilled holes—160 to 80
- Reamed holes—140 to 50
- Broached holes—125 to 40

Note: Finish marks (symbols) are *not* placed on drilled, reamed, or broached holes, even though these processes actually finish these surfaces.

FINISH MARKS: IN PRACTICE

- Finish marks are placed on the edge view of the surface that is to be finished.
- Finish marks are placed on all edges to be finished.

Example: How many finished surfaces are in the object below?

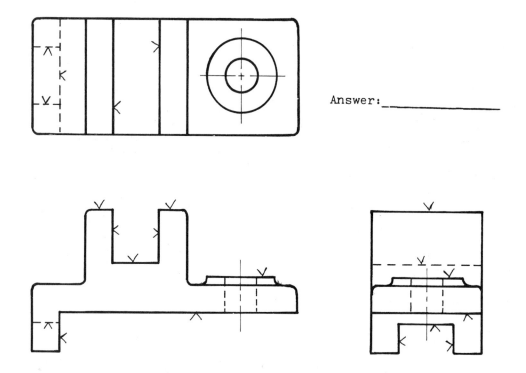

Answer:_____

(There are 11 finished surfaces on this part. Each surface is counted only once.)

The abbreviation F.A.O. indicates the part is to be *"finished all over."*

MISCELLANEOUS PRACTICE WORKSHEET

See end of chapter for answers.

1. List four kinds of *'keys'* used to prevent the rotation of a wheel on a shaft.
    _____
    _____

2. What is a "reference" dimension? _____
    _____
    _____

3. List the two kinds of knurls used in industry. _____
    _____

4. Which knurl has the most ridges per inch, a knurl with a 14 pitch or one
    with a 33 pitch? _____
    _____

5. Why is an undercut or relief used? _____
    _____

6. Is a #10, round head screw, a 'machine' or 'cap' screw? _____
    _____
    _____

7. List five kinds of *cap screws.* _____
    _____
    _____

8. What is a microinch? _____
    _____

9. How many microinches are *average* smooth? _____
    _____

10. How smooth a finish can be expected from a sandcasting? _____
    _____

STANDARD PRACTICES

In industry, the top limit is considered the most important figure, thus, the machinist strives to make the part that size.

In calling-off a hole, the *smallest* limit is placed on top. The machinist tries to make the hole that size. If the hold comes out too small it can be redrilled larger. If it is drilled a little larger than the top limit, it probably will still fall within the bottom, or *largest* limit.

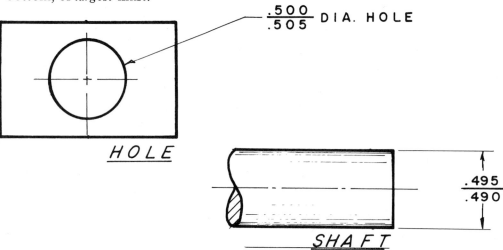

In calling-off a shaft, the *largest* limit is placed on top. The machinest tries to make the shaft that size. If the shaft comes out larger it can be re-machined smaller. If it is machined a little smaller than the top limit, it probably will still fall within the bottom, or *smallest* limit.

M.M.C. (Maximum Material Condition)

If an object was placed on a scale with a hole size of .500, it would weigh *more* than an object with a hole size of .505, thus, this is maximum material condition.

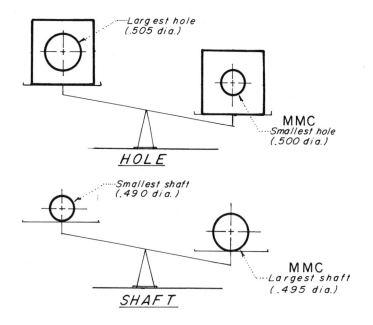

If a shaft was placed on a scale, the shaft dia. of .495 would weigh *more* than a shaft dia. of .490, thus, this is maximum material condition.

ALLOWANCE

Allowance is the difference in size of mating parts at *Maximum* Material Condition (M.M.C.)

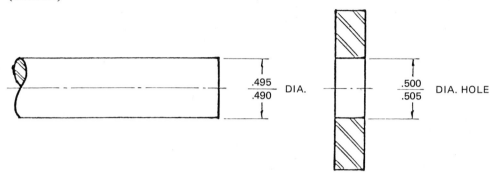

The *largest* shaft allowed is .495 dia. (MMC)

The *smallest* hole allowed is .500 dia. (MMC)

Thus,   .500 *smallest* hole (MMC)
        − .495 *largest* shaft (MMC)
        = .005 *allowance* between parts.

CLEARANCE

Clearance is the difference in size of mating parts at *Minimum* Material Condition (M.M.C.)

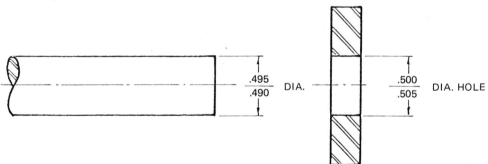

The *smallest* shaft allowed is .490 dia.

The *largest* hole allowed is .505 dia.

Thus, .505 *largest* hole
      −.490 *smallest* shaft
      =.015 *clearance* between parts

NOMINAL SIZE

Nominal size is a size used for general identification only. A steel plate that is referred to as being 1/4 inch (6mm) thick, if measured is actually more or less than 1/4 inch (6mm) thick thus, the 1/4 inch (6mm) is used only for *general identification* of the steel plate.

STANDARD PRACTICES WORKSHEET

Using drawing A-196531 answer the following questions in the spaces provided. See end of chapter for answers.

1. What would the closest metric equivalent of the thread at 'A'? (Use the standards chart in this chapter.) _____

2. What would the "reference" dimension 'B' equal? _____

3. What does the hidden line at 'C' indicate? _____

4. Calculate dimension 'D'. _____

5. What is the surface finish on surface 'E'? _____

6. What is the diameter at 'F'? _____

7. What is the surface finish on surface 'G'? _____

8. What is the distance from surface 'H' to surface 'E'? _____

9. What does 'I' indicate? _____

10. What is the diameter at 'J'? _____

11. What size chamfer is used on diameter 'I'? _____

12. What is the size of the *undercut* on diameter 'J'? _____

13. How deep do the *'full threads'* of the 3/8-16 UNC threaded hole go into the part? _____

14. What size round stock would probably be used to make this part? _____

15. How wide is the keyway? _____

16. Is the knurl portion, fine, medium, or coarse? _____

17. What surface finish is used on the O.D.? _____

18. What is the part made of? _____

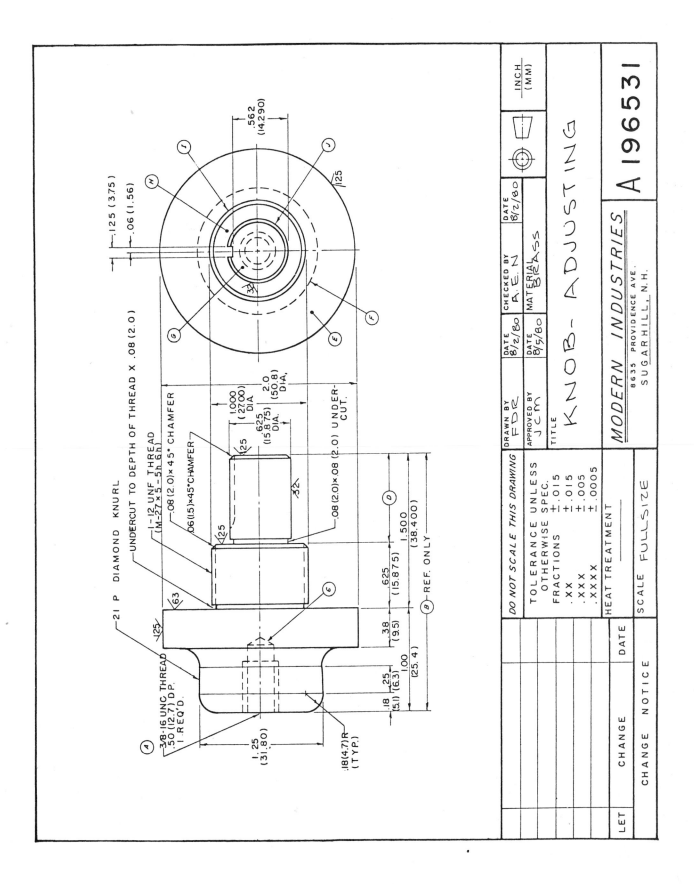

FUSION WELDING

In *fusion welding*, a welding rod is melted and combined with the metal parts that are to be fastened together. The parts will be permanently joined after cooling. The process can be done using torches or high electric power.

TYPES OF TYPICAL JOINTS

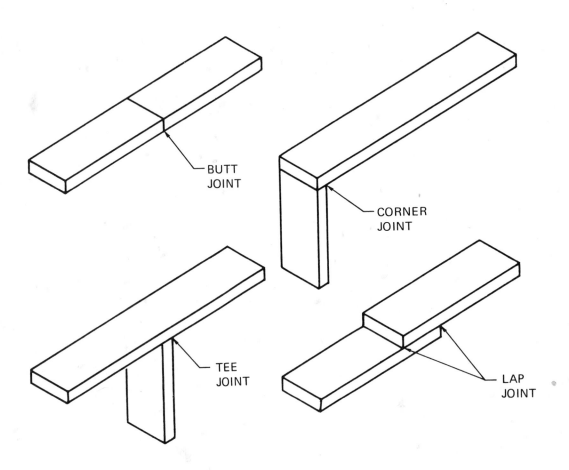

SYMBOLS

Indicated below are the symbols for each type of weld.

TYPE OF WELDS

① BACK OR BACKING WELD	② FILLET WELD	③ PLUG WELD	④ SQUARE WELD	⑤ V WELD	⑥ BEVEL WELD	⑦ U WELD	⑧ J WELD
⌒	◺	▭	‖	∨	⋁	∪	⋃

Weld symbols

Below is an example of each type of weld.

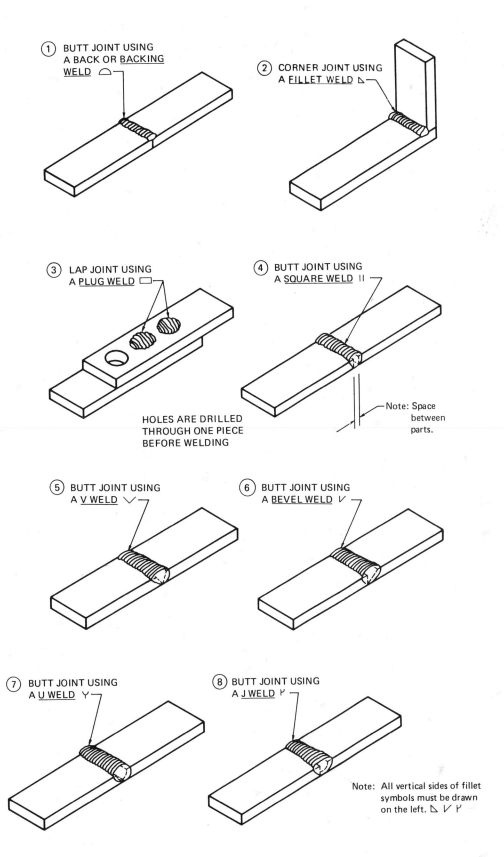

① BUTT JOINT USING
A BACK OR BACKING
WELD ⌓

② CORNER JOINT USING
A FILLET WELD ⊿

③ LAP JOINT USING
A PLUG WELD ▭

HOLES ARE DRILLED
THROUGH ONE PIECE
BEFORE WELDING

④ BUTT JOINT USING
A SQUARE WELD ‖

Note: Space
between
parts.

⑤ BUTT JOINT USING
A V WELD ∨

⑥ BUTT JOINT USING
A BEVEL WELD ⋁

⑦ BUTT JOINT USING
A U WELD ⋃

⑧ BUTT JOINT USING
A J WELD ⌐

Note: All vertical sides of fillet
symbols must be drawn
on the left. ⊿ ⋁ ⌐

PLACING WELD SYMBOLS

Rule 1. When the weld symbol is placed below the reference line, (B), the weld appears on the same side as the arrowhead.

Rule 2. When the weld symbol is placed above the reference line (A), the weld appears on the opposite side of the arrowhead.

Rule 3. When the weld symbol is placed above and below the reference line (A and B), the weld appears on both sides of the arrowhead.

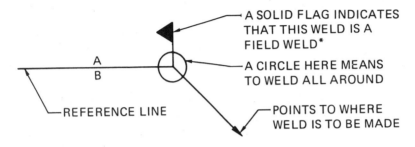

Rules 1, 2, and 3 use a filler weld symbol.

Usually material over .125 inch (3) thick requires a *groove* (square, V, beveled, U, or J). Using the basic weld symbol, the arrowhead is pointed toward the part that has the groove.

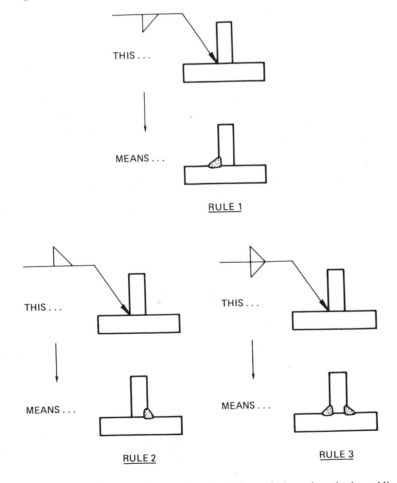

*A field weld symbol indicates that the weld must be made at the work site and not in the welding shop.

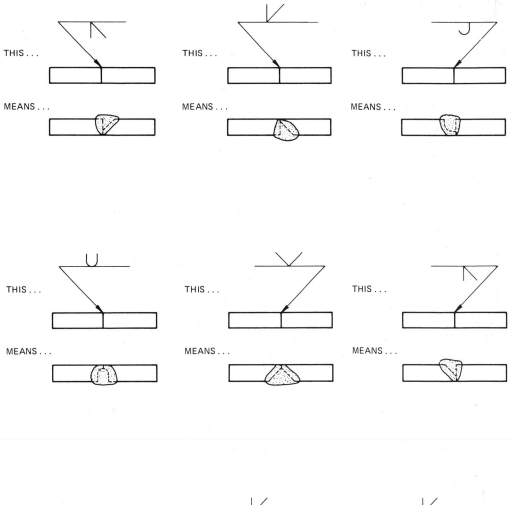

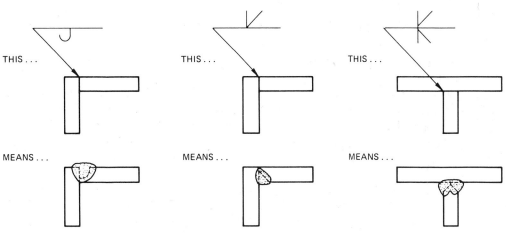

Groove welds shown require that the groove be machined to a specific size conforming to a specific delineation.

WELDING SYMBOLS WORKSHEET

Below each weld callout, sketch the position and shape of the weld called for. See end of chapter for answers.

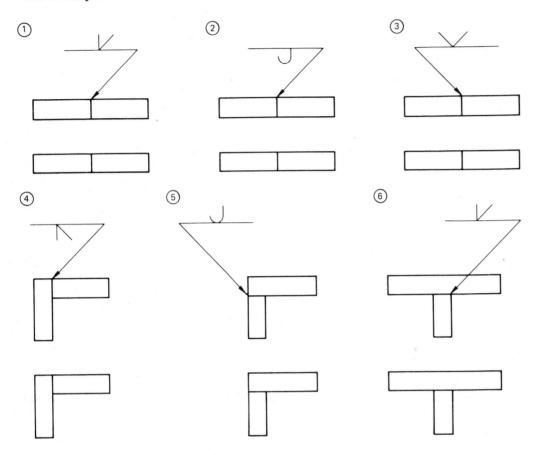

REFERENCE LINE NOTATIONS

There are various *notations* placed on or around the reference line. Below are a few of the more widely used standard notations. Each tells the welder exactly how the drafter wants the part(s) welded. The .25 X .375 inch (6 X 10) *size notation* means the weld is approximately .25 X .375 inch and is welded the whole length of the part.

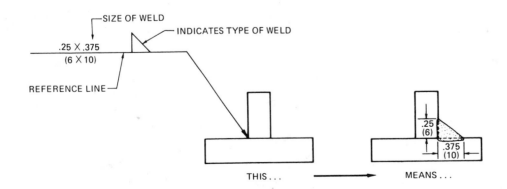

In the illustration below, two more notations are added. The first means the length of each weld, and the second means the distance from center to center of each weld or pitch. *Pitch* refers to the distance, center to center, of each weld. The 2-4 notation means that the weld is to be 2 inches long with a center-to-center distance of 4 inches. The notations would indicate millimeters if the metric system is used.

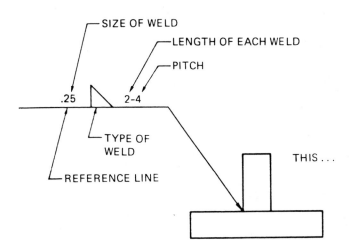

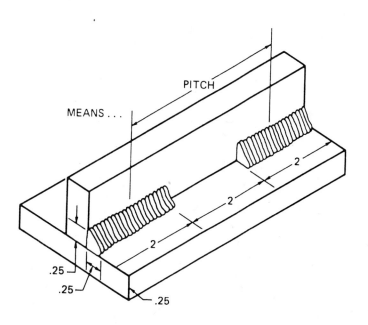

BASIC WELDING WORKSHEET I

On the reference line, draw the welding symbol required to obtain the weld and spacing shown on the isometric drawings given. See end of chapter for answers.

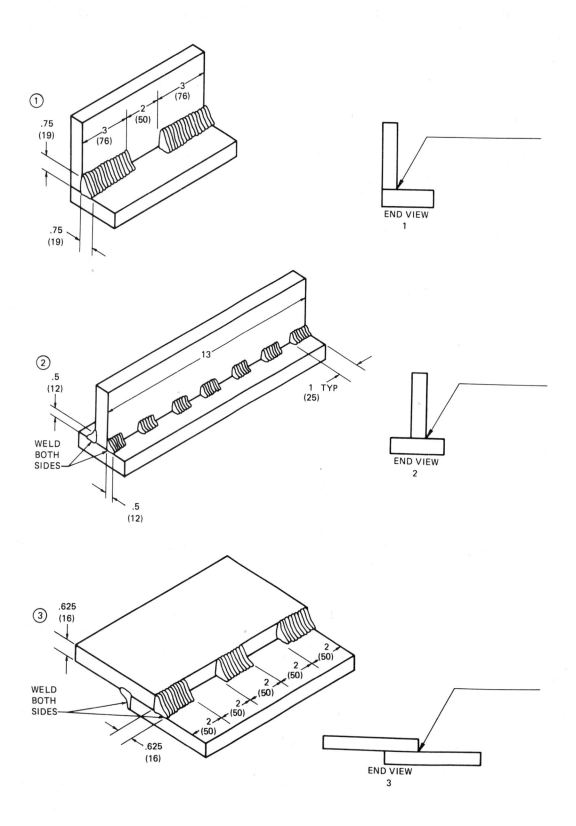

RESISTANCE WELDING

Resistance welding is the process of passing an electric current through a spot where the parts are to be joined. Symbols for resistance welding are shown below.

TYPE OF WELDS

① RESISTANCE–SPOT WELD	② PROJECTION WELD	③ RESISTANCE–SEAM WELD

Note that the symbols in the illustration below are similar to those used in fusion welding. As with fusion welding, notations are done on the side of the arrow where they appear.

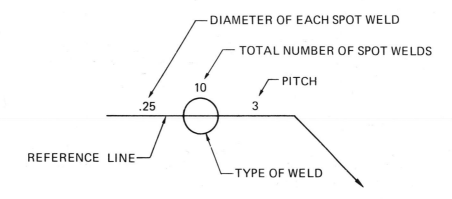

SPOT WELDING

Spot welding joins parts together with small circles or spots of heat. The illustration below shows how a drafter would draw and dimension a drawing.

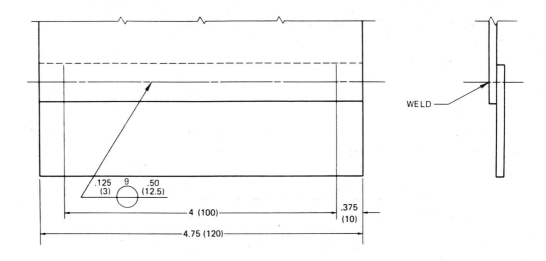

How the part would be spot welded:

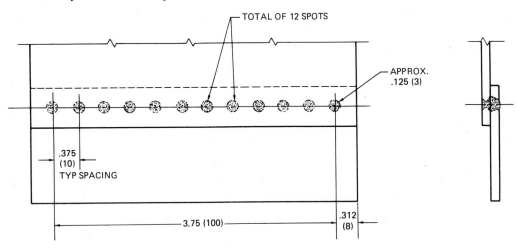

FLUSH SYMBOL

A *flush symbol* is used to indicate that one or both surfaces must be ground smooth:

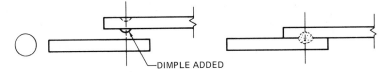

PROJECTION WELD

A *projection weld* is the same as a spot weld except one part has a *dimple* stamped into it at each spot where it is to be welded. This dimple allows more penetration and, as a result, is a better weld:

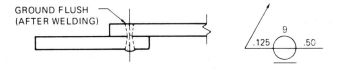

The illustration below shows how a drafter would draw and dimension a drawing. Note that the symbol is located below the reference line indicating that the dimple is on the part that is on the arrow side.

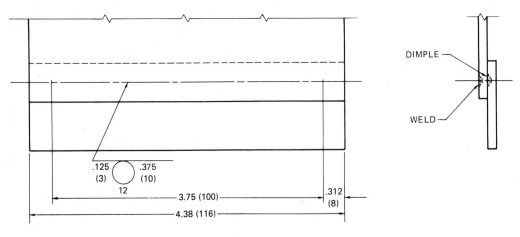

How the welder would weld the project:

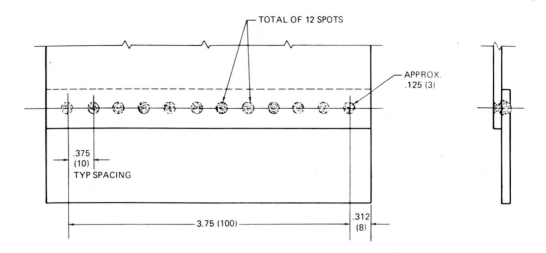

RESISTANCE SEAM WELD

A *resistance seam weld* is like the spot weld process except the weld is continuous from start to finish. The illustration below shows how the drafter would draw and dimension a drawing:

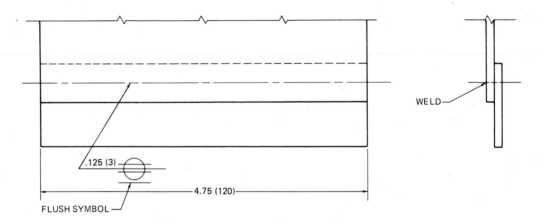

The illustration below shows how the welder would seam weld the project:

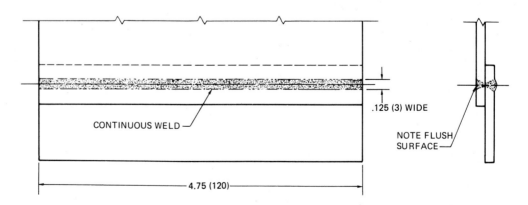

BASIC WELDING WORKSHEET II

Study each isometric sketch. On the reference lines, place the information needed to obtain the desired welds illustrated at the left of each problem. See end of chapter for answers.

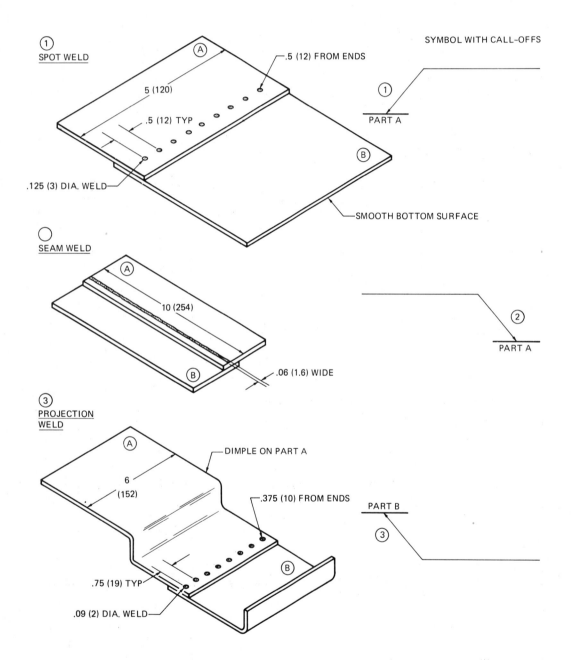

① SPOT WELD

.5 (12) FROM ENDS

5 (120)

.5 (12) TYP

.125 (3) DIA. WELD

SMOOTH BOTTOM SURFACE

SYMBOL WITH CALL-OFFS

① PART A

SEAM WELD

10 (254)

.06 (1.6) WIDE

② PART A

③ PROJECTION WELD

DIMPLE ON PART A

6 (152)

.375 (10) FROM ENDS

.75 (19) TYP

.09 (2) DIA. WELD

PART B

③

METALLURGY

A craftsperson should know about the behavior, characteristics, and properties of metals. These pages give a very general working knowledge of metals. the-job study must be done by the drafter.

Pure metals by themselves are usually too soft and weak to be used for machine parts. Thus, alloys are used. An *alloy* is simply a mixture of metals and chemical elements.

Materials are carefully chosen to give the best working life of the part to be made and still be in line cost-wise with competition. In industry, most companies have one or more metallurgists who work with the engineering department to assist in the selection of correct metal or alloy for the design and function of each machine part. *Metallurgy* is the art and science of separating metals from their ores and preparing them for use.

CHARACTERISTICS OF METALS AND ALLOYS

The composition of metal and various chemical elements regulates the mechanical, chemical, and electrical properties of that metal. The following terms describe certain characteristics and capabilities associated with metals and alloys:

Strength is the ability to resist deformation.

Plasticity is the ability to withstand deformation without breaking. Usually hardened metals have strength but are very low in plasticity. They are brittle.

Ductility describes how well a material can be drawn out. This is an especially important characteristic for wire drawing and metal shape forming.

Malleability is the ability of a metal to be shaped by hammering or rolling.

Elasticity is the ability of metal to be stretched and then return to its original size.

Brittleness is a characteristic of metal to break with little deformation.

Toughness describes a metal that has high strength and malleability.

Fatigue limit is the stress, measured in pounds per square inch, at which a metal will break after a certain number of repeated applications of a load has been applied.

Conductivity describes how well a metal transmits electricity or heat.

Corrosion resistance describes how well a metal resists rust. Note that *rust* adds weight, reduces strength, and ruins the overall appearance of a metal.

HEAT TREATMENT

Heat treatment of metals and alloys provides certain desirable properties. Listed are a few basic terms associated with heating treatment:

Aging is a process that takes place slowly at room temperature.

Quenching is the process of cooling a hot metal in water or oil. In special cases it can be quenched in sand, lime, or asbestos rather than liquid in order to slow down the cooling process.

Tempering or *"drawing"* is the process of reheating and cooling by air. Tempering increases toughness, decreases hardness, relieves stress, and removes some brittleness.

Annealing is the process of softening a hardened metal so it can be shaped or machined. Annealing also removes internal stresses which cause warping and other deformations.

Case hardening hardens only the outer layer of the material. In this process, the outer layer of the metal absorbs carbon or nitrogen, thus hardening the outer layer.

Hardness testing is done by two methods: the Brinell hardness test or the Rockwell hardness test. The *Brinell method* uses a hardened steel ball that is pressed into the surface of the metal under a given pressure or load. The depth the ball goes into the metal surface is measured by a microscope and is converted into a hardness reading. The *Rockwell method* is very much the same except the hardness value is read directly from a scale attached to the tester.

DRAWING NOTATIONS

It is important that any part to be hardened is noted on the drawing. This notation must include the material, the heat-treating process, and the hardness test method and number.

TYPES OF METALS

There are two classifications of metal: (1) *ferrous,* or those that contain iron; and (2) *nonferrous,* or those that do not contain iron.

Ferrous Metals

Cast iron is widely used for machine parts. It is relatively inexpensive and easily cast into most any shape. It is a hard metal, strong, and has a good wear-ability. It responds very easily to almost all heat treating processes but tends to be brittle. *Malleable iron* is used where parts are subject to shock.

Steel is an alloy composed of iron and other chemical elements. It is important to remember that carbon content in steel regulates the properties of various types of steel. There are four classes of steel: carbon, alloy, stainless, and tool. A standard system of designating steel has been established by the American Iron and Steel Institute (A.I.S.I.) and the Society of Automotive Engineers (S.A.E.) to describe the type of steel to be used. SAE1010 steel, for example indicates carbon steel with approximately 0.10% carbon.

PROPERTIES, GRADE NUMBERS & USAGES			
Class of Steel	*Grade Number	Properties	Uses
Carbon - Mild 0.3% carbon	10xx	Tough - Less Strength	Rivets - Hooks - Chains - Shafts - Pressed Steel Products
Carbon - Medium 0.3% to 0.6% carbon	10xx	Tough & Strong	Gears - Shafts - Studs - Various Machine Parts
Carbon - Hard 1.6% to 1.7%	10xx	Less Tough - Much Harder	Drills - Knives - Saws
Nickel	20xx	Tough & Strong	Axles - Connecting Rods - Crank Shafts
Nickel Chromium	30xx	Tough & Strong	Rings Gears - Shafts - Piston Pins - Bolts - Studs - Screws
Molybdenum	40xx	Very Strong	Forgings - Shafts - Gears - Cams
Chromium	50xx	Hard W/Strength & Toughness	Ball Bearings - Roller Bearing - Springs - Gears - Shafts
Chromium Vanadium	60xx	Hard & Strong	Shafts - Axles -Gears - Dies - Punches - Drills
Chromium Nickel Stainless	60xx	Rust Resistance	Food Containers - Medical/Dental Surgical Instruments
Silicon - Manganese	90xx	Springiness	Large Springs

*The first two numbers indicate type of steel, the last two numbers indicate the approx. average carbon content — 1010 steel indicates, carbon steel w/approx. 0.10% carbon.

Nonferrous Metals

Copper is soft, tough, and ductile. It is a good conductor of both electricity and heat.

Brass is an alloy of copper (copper/zinc) and very workable, tough, and ductile.

Bronze is another alloy of copper (copper/tin). It is a serviceable, strong, and tough metal.

Aluminum is very malleable, ductile, and a good conductor of electricity and heat. It is very light in weight. Aluminum cannot be heat treated; thus, to increase its hardness, other alloys and elements must be added.

Magnesium is perhaps the lightest metal used today. It is a good conductor of electricity and heat, nonmagnetic, easily machined, but highly inflammable while machining.

SHAPES OF METALS

Metals are purchased from the manufacturer in standard shapes and sizes. The drafter must use the correct call outs when indicating what material is to be used. Methods of designating measurements of basic shapes are given below.

Extrusion is one method of forming very odd or special shapes, similar to squeezing toothpaste from a tube. The round opening is like a die (the required shape) and the toothpaste represents the metal to be shaped.

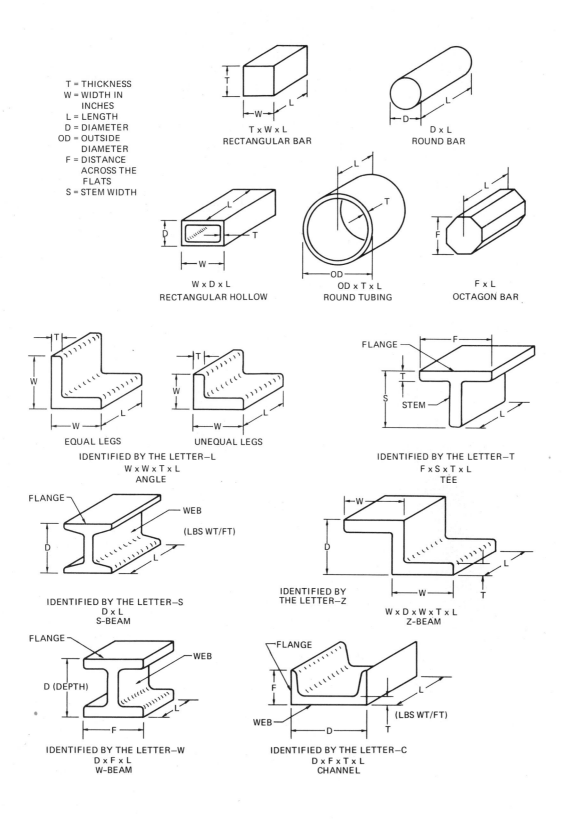

T = THICKNESS
W = WIDTH IN INCHES
L = LENGTH
D = DIAMETER
OD = OUTSIDE DIAMETER
F = DISTANCE ACROSS THE FLATS
S = STEM WIDTH

T x W x L
RECTANGULAR BAR

D x L
ROUND BAR

W x D x L
RECTANGULAR HOLLOW

OD x T x L
ROUND TUBING

F x L
OCTAGON BAR

EQUAL LEGS   UNEQUAL LEGS
IDENTIFIED BY THE LETTER—L
W x W x T x L
ANGLE

FLANGE   STEM
IDENTIFIED BY THE LETTER—T
F x S x T x L
TEE

FLANGE   WEB
(LBS WT/FT)
IDENTIFIED BY THE LETTER—S
D x L
S-BEAM

IDENTIFIED BY THE LETTER—Z
W x D x W x T x L
Z-BEAM

FLANGE   WEB
IDENTIFIED BY THE LETTER—W
D x F x L
W-BEAM

FLANGE   WEB
(LBS WT/FT)
IDENTIFIED BY THE LETTER—C
D x F x T x L
CHANNEL

WEIGHT OF MATERIALS

The table below indicates the average weight per cubic foot of certain materials.

WEIGHTS OF MATERIALS

Material	Avg. Lbs. per Cu. Ft.	Avg. Kg. per Cu. Meter	Material	Avg. Lbs. per Cu. Ft.	Avg. Kg. per Cu. Meter
Aluminum	167.1	2676	Mahogany, Honduras, dry	35	564
Brass, cast	519	8296	Manganese	465	7448
Brass, rolled	527	8437	Masonry, granite or		
Brick, common and			limestone	165	2648
hard	125	2012	Nickel, rolled	541	8649
Bronze, copper 8, tin 1	546	8754	Oak, live, perfectly dry		
Cement, Portland, 376 lbs.			.88 to 1.02	59.3	953
net per bbl	110–115	1765–1836	Pine, white, perfectly dry	25	388
Concrete, conglomerate,			Pine, yellow, southern dry	45	706
with Portland cement	150	2400	Plastics, molded	74–137	1200–2187
Copper, cast	542	8684	Rubber, manufactured	95	1518
Copper, rolled	555	8896	Slate, granulated	95	1518
Fibre, hard	87	1377	Snow, freshly fallen	5–15	70–247
Fir, Douglas	31	494	Spruce, dry	29	459
Glass, window or plate	162	2577	Steel	489.6	7837
Gravel, round	100–125	1586–2012	Tin, cast	459	7342
Iron, cast	450	7201	Walnut, black, perfectly dry	38	600
Iron, wrought	480	7695	Water, distilled or pure rain	62.4	988
Lead, commercial	710	11,367	Zinc or spelter, cast	443	7095

MACHINE TOOL OPERATIONS

The drawing must include all information required so a skilled craftsperson can manufacture a finished part from raw stock. Basic manufacturing processes shape raw stock by:

- *Cutting* into shape
- *Molding* into shape by casting or machine press
- *Pounding* into shape by forging
- *Forcing* into shape by bending
- *Fabricating* into shape, using parts manufactured from a combination of the above processes, by welding, riveting, screwing, or nailing the parts together

Machines that cut metal are called *machine tools*. There are over 400 kinds of machine tools, each designed to do a specific operation. All machine tool operations can be divided into five basic processes: Drilling, turning, planing, milling, and grinding.

A reader does not have to be machinist but should have a basic knowledge of machine tool operations in order to dimension drawings and "talk the language" of the skilled machinist.

South St. Paul Public Library
106 Third Avenue North
South St. Paul, MN 55075

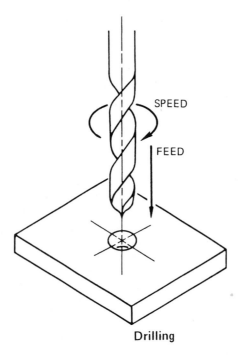

Drilling

Drilling

Drilling is probably the most basic of all machine tool operations. The process is done with a rotating tool called a *drill,* (see illustration). The drill is held by a *chuck* and is rotated and fed into the part to be drilled. Holes are drilled before boring, reaming, countersinking, or counterboring operations can be completed.

Turning

Turning is the process of rotating the part that is to be machined and carefully pressing a cutting tool against it as it rotates, (see illustration). A *lathe* is used for turning down stock and can be used for other machine operations such as drilling, boring, threading, cutting, milling, grinding, and knurling. A *turret lathe* is a lathe with a six-sided tool holder, called a *turret,* to which various cutting tools are attached. This attachment enables the lathe to do many operations without resetting the tools once each of the tools in the turret has been set.

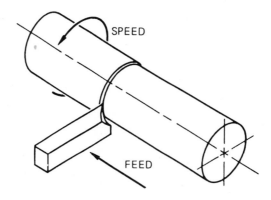

Turning

Planing

Planing is the process of shaving material from raw stock very much like a carpenter does with a simple hand plane. The major difference is that the plane or cutting edge is stationary, and the part that is to be planed is moved back and forth, (see illustration). A *shaper* is like a planer except, in shaping, the plane or cutting edge moves and the part is stationary. A *broach,* which is generally used to cut key slots and similar configurations, falls into the category of a planer in the way it operates. These tools can be worked horizontally, vertically, or angularly.

Grinding

Grinding is a machine operation where the part is brought into contact with a rotating abrasive wheel, (see illustration). With this process, it is possible to obtain very close, precise tolerances. Grinders that finish round parts are called *cylindrical grinders.* Those that grind flat parts are called *surface grinders.* Those that grind holes are called *internal grinders.*

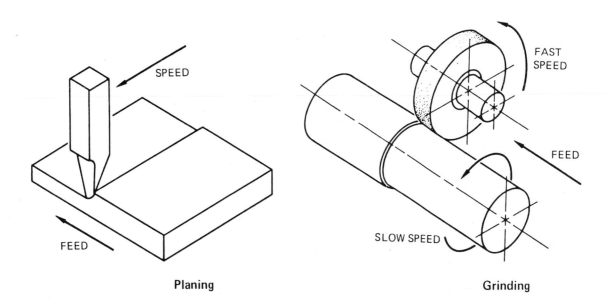

Planing   Grinding

Milling

Milling is the process of bringing the part into contact with a rotating cutting tool having many edges, (see illustration). The shape of the cutting edges are similar to a woodworking circular saw blade except they are usually wider and have a greater variety of shapes, (see illustrations). Milling machines produce cuts that are flat, round, sharp, and a combination of these shapes. Common milling machine processes are cutting slots and grooves, cutting gear teeth, making threads, boring holes, and rounding corners of parts.

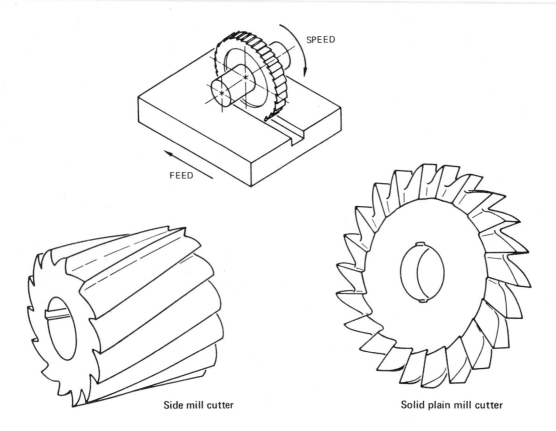

Side mill cutter   Solid plain mill cutter

CASTINGS

Casting is the process of forming metal parts to rough size and shape by pouring molten metal into a mold. This process is similar to the way ice cubes are formed by pouring water into a tray and freezing it.

There are many forms of casting, varying in techniques and precision. To explain casting, one basic method—sand casting—is illustrated.

The patternmaker constructs the pattern of the object to be cast. If the object, such as a bookend, has one flat side, the pattern is a one-piece pattern. If the object to be cast is round, a two-piece *split pattern* is used (see illustration).

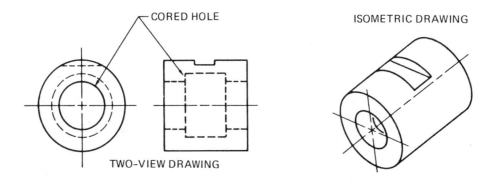

Object to be cast

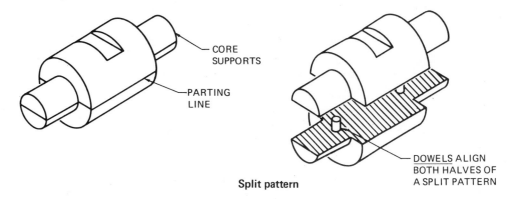

Split pattern

If the object has a large hole through it, a *core support* is located inside the mold, (see illustration). This prevents molten metal from solidifying in the hole during the casting process. When the core is removed, a rough cored hole remains which is later machined to the correct size. No cores are used on small holes as they are simply drilled into the finished casting.

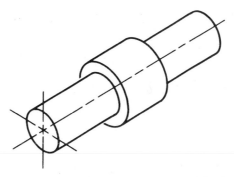

Core made from baked sand and held together with a bonding agent

The *flask* is a hollow box with no top or bottom that holds the sand and the mold. The *cope* is the top half of the flask and the *drag* is the lower half of the flask. A *socket* aligns the cope and drag. The flask is placed on a *molding board.*

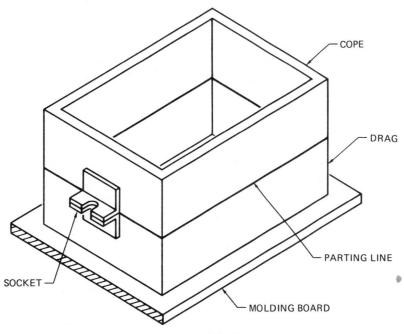

A flask

How to Make a Metal Casting

The following description of a sand casting is a general description only and is not intended to be used in actual casting processes.

Step 1. The lower half of the split pattern is placed on the molding board. The drag is centered around the pattern.

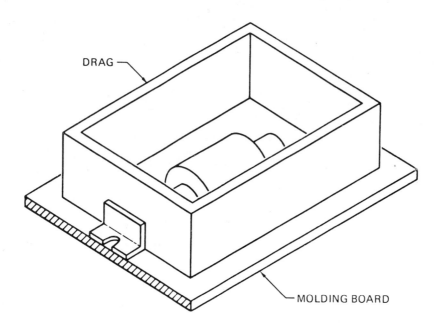

Step 2. The drag is filled with sifted sand which is packed firmly around the split pattern and leveled off. Another molding board is placed on top to hold the sand in place, and the drag is turned over.

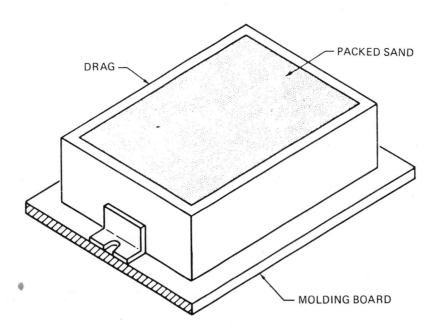

Step 3. The molding board is removed from the top to expose the pattern.

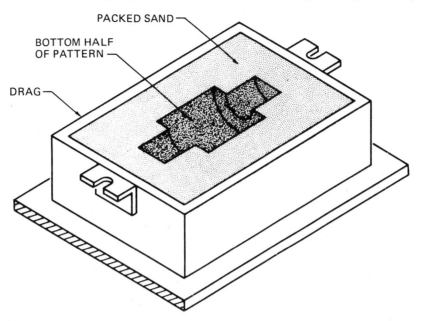

Step 4. The dowel pins are put in place and the top of the split pattern is positioned on the lower half. The cope is then placed on the drag, locked into position, and filled with sand. The sand is tightly packed around the pattern.

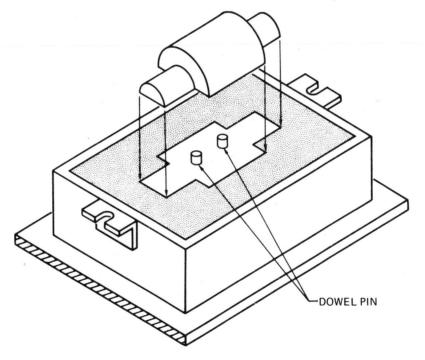

Step 5. Molten metal is poured into the mold through the *sprue hole.* The gases escape through the *riser hole* during the casting process. These holes are made while the pattern is in place so that the sand from the cope will not be forced into the hollow mold. An alternate method is to locate the sprue and riser holes to the left and right of the pattern and, with the flask apart, cut a groove leading from them to the mold.

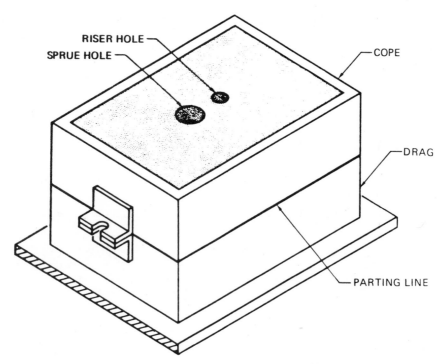

RISER HOLE
SPRUE HOLE
COPE
DRAG
PARTING LINE

Step 6. After cutting the sprue and riser holes, the cope is carefully removed from the drag and set aside, and the pattern removed from the drag. This leaves the top half of the mold in the cope and the bottom half of the mold in the drag. Round patterns will lift easily from the sand. Flat patterns tend to stick and can be damaged when removed. To prevent this, flat patterns are tapered on their sides. This taper is called a *draft*. The angle of draft is shown on a casting drawing and varies with the kind of material being cast. Sharp corners on flat patterns are also rounded off slightly for the same reason.

The core is put in place and the cope returned to the drag and locked into position. The casting is now ready to be poured. After the molten metal has solidified, the casting is removed from the sand and the core is removed from the casting. The casting is now ready to be machined.

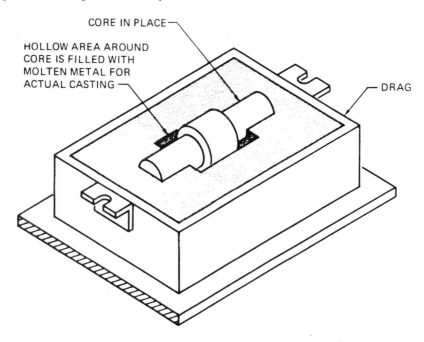

CORE IN PLACE
HOLLOW AREA AROUND
CORE IS FILLED WITH
MOLTEN METAL FOR
ACTUAL CASTING
DRAG

Section View of a Casting

A complex casting is drawn in two sections: One shows the pattern and the other shows the machining of the casting.

A section view of a complete flask with the core in place is shown below. The molten metal is poured into the funnel-shaped sprue hole until the metal fills the cavity and comes out the riser. The riser allows air to escape while the mold is filled and feeds the casting while it cools.

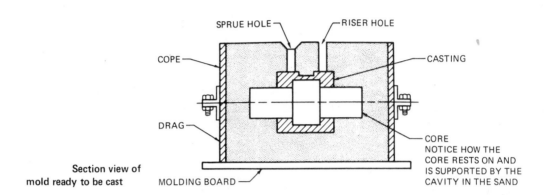

Section view of mold ready to be cast

The sand core is easily broken up and removed, leaving a cavity inside the casting. When the casting is removed from the sand, the sprue and riser are still attached.

The sprue and riser are easily removed by breaking or cutting them off. They are then hand-ground. A sand casting is rough and the critical surfaces must be smoothed and machined to exact sizes.

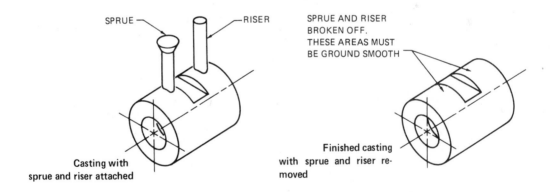

Rounds and Fillets

Any part that is formed by casting should be designed with rounds and fillets. *Rounds* are merely rounded outside corners. *Fillets* are rounded inside corners.

Rounds and fillets are used for three reasons:

1. Greater strength for inside corners (fillets)
2. Safer to handle; no sharp corners (rounds)
3. Gives the finished casting a much better appearance (rounds/fillets)

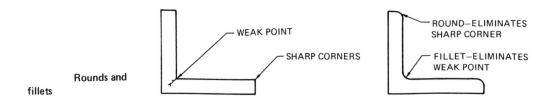

Bosses, Pads, and Machining Lugs

Bosses and *pads* serve the same function. They are raised surfaces that are machined to provide a smooth surface for mating parts. This method of designating saves material and machining time.

A *boss* is a round, raised surface. A *pad* can be any shape raised surface. Usually the bottom surface of a part is machined to provide a smooth, solid surface to support the part. The rest of the sand casting's surface is very rough. Proper use of rounds and fillets are illustrated below.

A *machining lug* is an extension of a surface to be machined. It is used for holding the casting because of its shape while machining. It too is removed after machining.

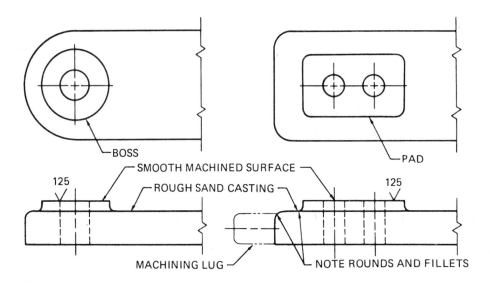

Boss, pad, and machining lug

Ribs and Webs

Ribs and webs are similar and often confused. Think of a *rib* as a member that supports other members. A *web* simply connects various members together. In designing castings, the general rule is to try to make all ribs and webs the same thickness. Otherwise, when the molten metal cools, thicker members cool last and tend to create warping and internal stresses.

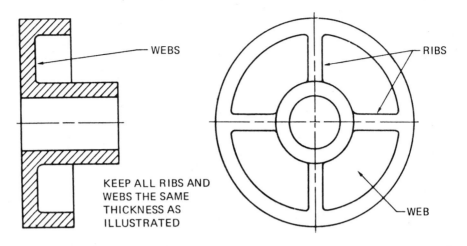

WEBS

RIBS

KEEP ALL RIBS AND
WEBS THE SAME
THICKNESS AS
ILLUSTRATED

WEB

Ribs and webs

Lug

A *lug,* sometimes referred to as an *ear,* is an extension added to the main part of the object. It usually holds parts together. The thickness of the lug should be approximately the same thickness as the round body thickness.

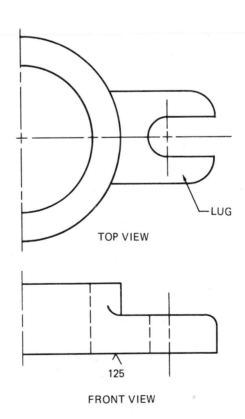

LUG

TOP VIEW

125

FRONT VIEW

Lugs

Shrink Rule

Patterns must be made larger than the desired size of the casting to compensate for the shrinkage which occurs when metal cools. A *shrink rule* is used by the patternmaker to overcome this difference. Different shrink rules are used for different materials.

Shrinkage allowances have been established for various metals:

- Cast iron and malleable iron—1/8 inch per foot (10 mm per meter)
- Copper, aluminum, and bronze—3/16 inch per foot (16 mm per meter)
- Steel—1/4 inch per foot (21 mm per meter)
- Lead—5/16 inch per foot (26 mm per meter)

The patternmaker uses the shrink-rule measurement which correspond to the shrinkage allowance. In this way, the pattern will be large enough to compensate for shrinkage, and the final casting will shrink to the desired original size.

MANUFACTURING PROCESSES WORKSHEET

See end of chapter for answers.

1. List three kinds of heat-treating methods and briefly explain each.

   _____
   _____
   _____
   _____

2. Why is an alloy used in place of a pure metal?

   _____
   _____
   _____
   _____

3. Explain the difference between ferrous and nonferrous metal.

   _____
   _____
   _____
   _____

4. What is fatigue limit?

   _____
   _____
   _____
   _____

5. Explain case hardening.

   _____
   _____
   _____
   _____

6. In a SAE 2010 steel callout, what do the first two numbers (20) indicate? The last two numbers (10)?

_____

_____

_____

_____

7. What are the five basic machine tool operations?

_____

_____

_____

_____

8. In using a drill, what does a chuck do?

_____

_____

_____

_____

9. What are the five manufacturing processes used to shape raw stock?

_____

_____

_____

_____

10. Explain shrinkage allowance.

_____

_____

_____

_____

11. What is the difference between a pad and a boss?

_____

_____

_____

_____

12. List three functions of a riser.

_____

_____

_____

_____

13. What is the basic rule for designing ribs and webs?

_____

_____

_____

_____

14. What makes up the flask in the casting process?

_____

_____

_____

_____

EVALUATION

Answer the following questions in the space provided. See end of chapter for answers.

1. What is the standard thread profile used for all nuts and bolts manufactured in the U.S.? _____

_____

2. What tool is used to produce interior threads? _____

_____

3. List five kinds of permanent fasteners. _____

_____

4. On an external thread, what is considered the *minor* diameter? _____

_____

5. What is pitch? _____

_____

6. What is a blind hole? _____

_____

7. Exactly what does *1/4 – 20   UNC – 3B* mean? _____

_____

8. Exatly what does M-12 x 1.25 – 5H 6H? _____

9. List four kinds of keys. _____

_____

10. Referring to a *knurl,* what does pitch mean? _____

_____

11. What is a ref. dimension? _____

_____

12. List four kinds of *cap* screws. _____

_____

13. How many microinches are considered to be an *average* finish? _____

14. How deep would the surface irregularities be for a 250 microinch finish?

_____

15. What is the M.M.C. of a hole with a tolerance of .750/.755? _____

16. What is the *clearance* of two parts with a shaft dia. .795/.790 and a hole dia. .700/.705? _____

_____

17. List two methods to *weld* parts together. _____

_____

18. On a *spot* weld, what does a line under the weld symbol indicate? _____

_____

19. Explain what 'a', 'b', 'c', and 'd' means.

a. _____

b. _____

c. _____

d. _____

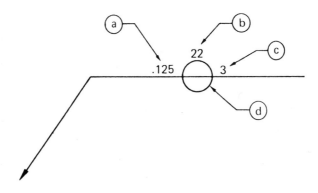

20. Using the illustration at the left, add all welding symbols with notations.

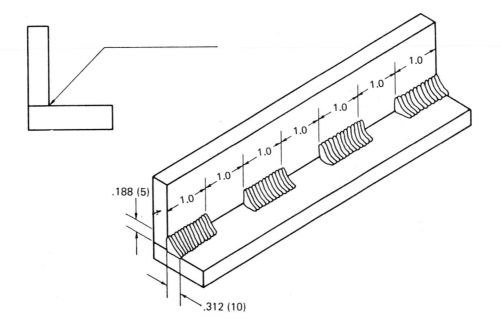

21. What is 'a' and 'b'—and what size are they? Answer 'c' and 'd'.

a. _____

b. _____

c. Are the threads coarse or fine? _____

d. Approx. how many threads will there be from top to bottom? _____

_____

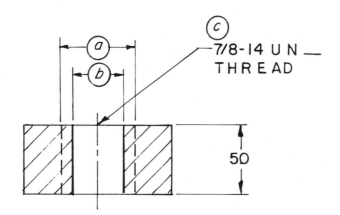

22. What does a line under a spot weld symbol indicate? _____
    _____

23. Where is the weld placed if the welding symbol is placed below the reference
    line? _____
    _____

24. What does a *solid* circle on the reference line of a fusion weld indicate?
    _____
    _____

25. What is the *allowance* of two parts with a shaft dia. .795/.790 and a hole
    dia. of .700/.705? _____

26. What is nominal size? _____
    _____

27. What is M.M.C.? _____

28. What surface would you expect to find for walls of an engine cylinder?
    _____

ANSWERS TO TERMS TO KNOW AND UNDERSTAND

Tap—Used to produce internal threads.

Die—Used to produce external threads.

Permanent Fasteners—Welds, brazes, staples, nails, glue, rivet.

Temporary Fasteners—Screws, bolts, keys, pins.

Purposes Of Threads—Fastening, adjusting, transmitting power, measuring.

Unified National Thread—A standard shaped (profile) thread (60°).

Square, Acme, Buttress, And Worm Threads—Various shapes or profile of threads
usually used to transmit power.

Major/Minor Dia. Of Threads—On external threads, major dia. is from crest to
crest; minor dia. is from root to root. On internal threads, major dia. is from root
to root; minor dia. from crest to crest.

Pitch—Distance from one thread to the exact same point on the next thread.

Thread Call-off—A standard method of describing a thread and/or a fastener.

T.P.I.—Threads per inch.

Keyway—A groove cut into a shaft and wheel, gear, etc. to hold the key in place.

Kinds Of Keys—Square, flat, woodruff, gib-head, Pratt & Whitney.

"Ref" Dimension—A dimension used *only* for reference that should *not* be used to manufacture the part.

Chamfer—A finish used on ends of shafts or rounded parts.

Knurling—The process of rolling depressions on a round surface.

Undercut (relief)—A cut made in a part to remove any thread run out or to remove any sharp radius in order to allow mating part to fit tightly against a shoulder.

Machine Screws—A flat, rounded, oval, or fillister head screw under a #12 in size.

Cap Screw—A flat, round, fillister, or hex, socket head screw over 1/4 inch in size.

Finish Marks—A code used to indicate surface finish requirements (measured in microinches).

M.M.C.—Maximum material condition.

Allowance—The dimensional difference between parts when at M.M.C.

Clearance—The dimensional difference between parts when at minimum material condition.

Nominal Size—Used for general identification only.

Types Of Welds—Fusion-resistant welding.

Types Of Joints—Butt, corner, and tee cap.

Symbol—

Alloy—A mixture of metals and chemical elements.

Metallurgy—Art and science of separating metals from their ores and preparing them for use.

Strength—Ability to resist deformation.

Plasticity—Ability to withstand deformation without breaking.

Ductility—Describes how well a material can be drawn out.

Malleability—A characteristic of metal that is shaped by hammering or rolling.

Elasticity—The ability of metal to stretch and return to its original size.

Brittleness—A characteristic of metal to break with little deformation.

Toughness—High strength.

Fatigue Limit—The stress (in lbs. per sq. in.) at which a metal will break after repeated loads have been applied.

Aging—A process that takes place slowly at room temperature.

Quenching—Cooling hot metal in oil or water.

Tempering—To reheat and cool slowly.

Annealing—The process of softening a hardened metal.

Case Hardening—Hardening only the outer layer of material.

Extrusion—A method of forming very odd or special shapes.

Drilling—Rotating drill is rotated or fed into the part.

Turning—Rotating the part and pressing a cutting tool into it as it rotates.

Planing—Process of shaving; similar to a carpenter's wood plane.

Grinding—Part is held against a rotating abrasive wheel.

Milling—Part is brought into contact with a cutting tool.

Lathe—A machine tool that, by a turning operation, can drill, bore, thread, cut, and mill.

Turret—Similar to a lathe except it grinds and knurls, has a six-sided tool holder, and can do many operations without resetting the tools.

Broach—Used to cut key slots; a type of planer.

Casting—Process of forming metal parts to rough size and shape by pouring molten metal into a mold.

Split Pattern—A two-piece pattern used for casting round objects.

Core Support—Inserted inside a mold to locate a large hole.

Molding Board—Board the split pattern is placed on.

Drag—Lower half of flask.

Cope—Upper half of flask.

Flask—The flask is made up of a cope, drag, and molding board.

Sprue—A funnel-shaped opening in the sand where the molten metal is poured.

Riser—Allows air to escape while filling the impression or hollow in the sand. It also indicates when the mold is full, and it feeds the casting while it is cooling.

Shrinkage Allowance—Metal shrinks when cooling, thus the wood pattern is made larger so the part will shrink to correct size.

Draft—A slight taper to the wood pattern so it can be pulled out of the sand.

Bosses/Pads—Raised areas on a part that are machined smooth. A boss is round; a pad is any other shape.

Lugs—Extend from the main part; usually used to assemble parts together.

THREADS WORKSHEET ANSWERS

1. Internal threads are usually provided by a *tap*.
2. Four kinds of temporary fasteners are: screws, bolts, keys, and pins.
3. Square, acme, buttress, and worm threads transmit power.
4. Six kinds of permanent fasteners are: welds, brazes, staples, nails, glue, and rivets.
5. *60°* is the standard angle which forms the unified national form thread.
6. 3/8 - 16 UNC - 2A, call-off.
7. #12 UNF has 28 T.P.I.
8. 9/16-18 UNF thread uses a 33/64 .5156 dia. tap drill.
9. The minor dia. of a 1/2 – 20 UNF is .435 (11.049).
10. A 7/8-14 UNF has a 22.225 mm diameter.
11. A 1/2 inch coarse thread has 13 T.P.I. thus, it would take 13 full turns to travel 1 inch.
12. The pitch of a 3/4 inch coarse thread is: .100 = 1 inch ÷ 10 T.P.I.

MISCELLANEOUS PRACTICE WORKSHEET ANSWERS

1. The four kinds of keys are, square, woodruff, gib-head, and Pratt & Whitney.
2. A 'reference' dimension is a dimension added to a drawing for *reference* only and should *not* be used in the manufacture of the part.
3. Two kinds of knurls are *straight* and *diamond* pattern.
4. 33 pitch has more ridges (14 pitch is *coarse,* 33 pitch is *fine*).
5. An undercut or thread relief is used to allow one part to fit *tightly* against another.
6. A *machine screw* is from #0 to 12 in size (*cap screws* are from 1/4 to 1 1/4 size).
7. Five kinds of caps crews are: flat, round, fillister, hex, and socket head.
8. A microinch = .000001.
9. 125 microinches is average.
10. A sandcasting has a finish of 300 to 1000 microinches.

Standard Practices Worksheet Answers

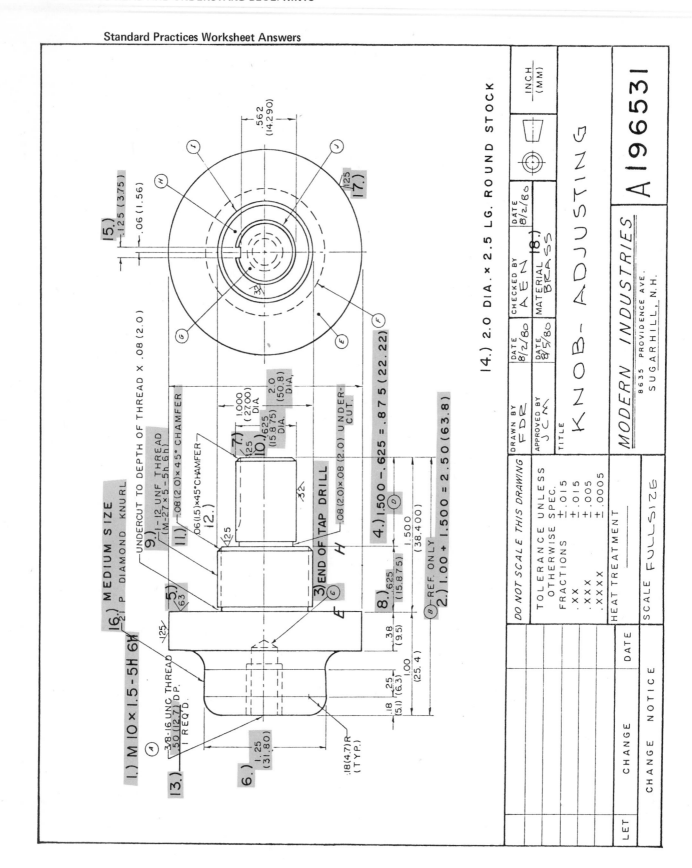

WELDING SYMBOLS WORKSHEET ANSWERS

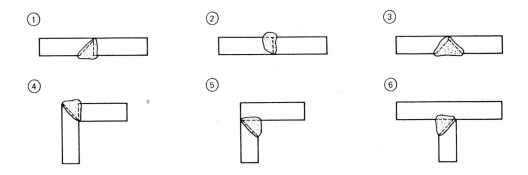

BASIC WELDING WORKSHEET I ANSWERS

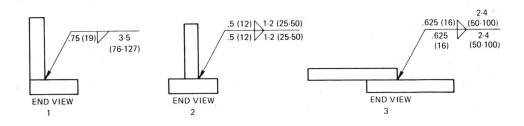

BASIC WELDING WORKSHEET II ANSWERS

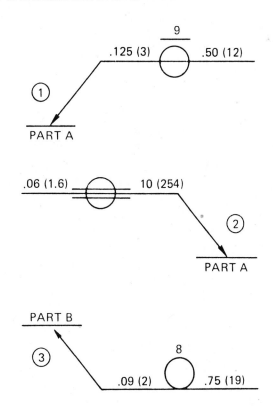

MANUFACTURING PROCESSES WORKSHEET ANSWERS

1. *Aging*—process which takes place slowly at room temperature.

   Quenching—process of cooling a hot metal in water or oil.

   Tempering or "drawing"—process of reheating and cooling by air. It increases toughness, decreases hardness, relieves stress, and removes some brittleness.

   Annealing—process of softening a hardening metal so it can be shaped or machined. It removes internal stresses which cause warping.

   Case hardening—hardens only the outer layer of the material.
2. Pure metals are usually too soft and very weak to be used by themselves.
3. Ferrous metal contains iron. Nonferrous metal does not contain iron.
4. The amount of stress, measured in pounds per square inch, at which a metal breaks after a certain number of loads is applied to it.
5. Case hardening is the process used to harden the outer surface or layer only. The outer surface absorbs carbon or nitrogen, thus hardening.
6. The first two didits (20) indicates the type of steel (nickel).
   The last two digits (10) indicates the approximate average carbon content—in this case, 0.10% carbon.
7. Drilling, turning, planing, grinding, and milling
8. A chuck holds the rotating drill in the drilling process.
9. Cutting, molding, pounding, forcing, and fabricating
10. Shrinkage allowance is the amount of material that is added to a pattern to compensate for shrinkage which occurs when the molten metal cools during the casting process.
11. A boss is a round, raised surface; a pad is any shape, raised surface
12. Allows air to escape during casting; indicates when the mold is full; and feeds the casting while it is cooling
13. Make all ribs and webs the same thickness.
14. Cope, drag, socket, molding board

EVALUATION ANSWERS

1. Unified national thread is the U.S. standard.
2. A *tap* produces interior threads.
3. Welding, brazing, stapling, nailing, gluing, and riveting are permanent methods to fasten parts together.
4. On an external thread, the minor dia. is measured from root to root.
5. Pitch is the distance from a point on one thread to the *exact* same point on the next thread.
6. A blind hole is a hole that does not go through an object.
7. 1/4 inch diameter/20 T.P.I./unified national coarse/tight fit/interior threads.
8. 11.963 mm dia./fine thread/pitch diameter tolerance of 5/crest diameter tolerance of 6/*interior* threads.
9. Square, flat, woodruff, gib-head, and Pratt & Whitney.
10. Pitch on a knurl means how many ridges per inch (25.4).
11. A ref (reference) dimension is a dimension added for reference *only,* and must *not* be used to manufacture the part.
12. Flat head, round head, fillister head, hex and socket head.
13. 125 microinches is average.

14. 250 x .000001 = .000250.
15. M.M.C. of a hole .750/.755 would be .750 (the smallest it can be).
16. .705 (largest hole) minus .790 (smallest shaft) = *.015 clearance* between parts.
17. Fusion welding and resistance welding.
18. A line under the weld symbol is the "flush" symbol, which indicates grind the weld smooth.
19. a. Dia. of each spot
    b. 22 spot weld points
    c. Spot welds 3 inches apart
    d. Spot weld symbol
20.

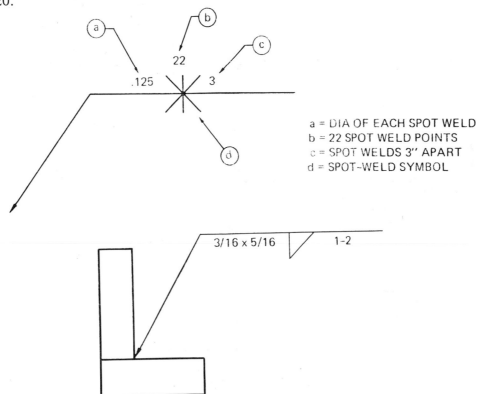

a = DIA OF EACH SPOT WELD
b = 22 SPOT WELD POINTS
c = SPOT WELDS 3" APART
d = SPOT-WELD SYMBOL

21. a. Major dia. (.875)
    b. Minor dia. (.782)
    c. Fine threads (14)
    d. 7 Full threads in .50 distance
22. A line under a welding symbols indicates a smooth surface in required after welding. (This is a 'flush' symbol).
23. When a welding symbol is placed *below* the reference line the weld is made on the *same side as the arrowhead.*
24. A solid circle indicates *field weld.*
25. .700, smallest hole (MMC) minus .795 largest shaft (MMC) = .005 *allowance* between parts.
26. Nominal size is for general identification only. The actual size may vary from the nominal size.
27. M.M.C. = Maximum material condition.
28. A extremely fine #8 finish is used on cylinder walls.

UNIT 8

TECHNICAL
INFORMATION

Objective: To learn how to use tabular drawings, interpret geometric tolerancing symbols, and read 1st angle projections.

TERMS TO KNOW AND UNDERSTAND

See end of chapter for answers.

Tabular Drawings _____

Datum Planes _____

"X" _____

"Y" _____

"Z" _____

Tabular Dimensions _____

Geometric Tolerance _____

Feature Control Note _____

T.I.R. _____

Ⓢ _____

Ⓜ _____

True Positioning _____

1st Angle Projection _____

TABULAR DRAWINGS

A tabular drawing is used by companies in order to save drafting time, as *one* drawing can serve to illustrate many similar parts. If a company manufactures various parts that have approximately the same size and shape, it can be drawn on a tabular drawing. A tabular drawing can be a detail drawing (individual parts) or an assembly drawing (many parts shown assembled). All variable dimensions are replaced by a reference letter; a table listing these letters with their corresponding dimensions is added below. In reading a tabular drawing care must be taken so as not to misread the table of reference dimensions.

For example, illustrated below is a 'U' bolt, the plan number is *A92544*. The 'U' bolts are all made of .375 (9.52) dia. rod, all threaded 3/8 x 16 UNC thread but the bend varies (radius A); the overall length varies, and the thread length varies. Note the numbering system used for each part (Drawing number dash 1, 2, 3, etc.)

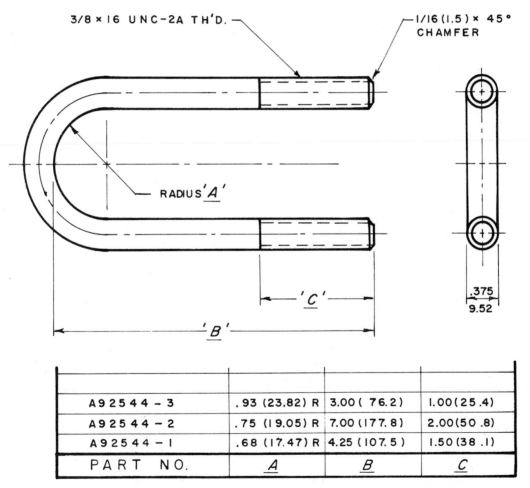

PART NO.	A	B	C
A92544 – 3	.93 (23.82) R	3.00 (76.2)	1.00 (25.4)
A92544 – 2	.75 (19.05) R	7.00 (177.8)	2.00 (50.8)
A92544 – 1	.68 (17.47) R	4.25 (107.5)	1.50 (38.1)

(PLAN NO. A92544)

TABULAR DRAWINGS WORKSHEET

Using drawing A22574-A, answer the following questions. See end of chapter for answers.

1. What scale is section A-A drawn to? _____

2. What is the I.D. (Dimension 'K') of part no. A22574-7? _____
   _____

3. What is the O.D. (diameter 'L') of part no. A22574-11? _____
   _____

4. Which part has the largest O.D.? _____

5. Surface 'M' in the front view is what surface in section A-A? _____
   _____

6. Surface 'J' in section A-A, is what surface in the right side view? _____
   _____

7. Surface 'D' in the right side view is what surface in section A-A? _____
   _____

8. What material is the part made of? How thick is it? _____
   _____

9. Which parts have 'B' and 'C' dimensions equal? _____

10. What is the I.D. (dimension 'K') of part no. A22574-10? _____
    _____

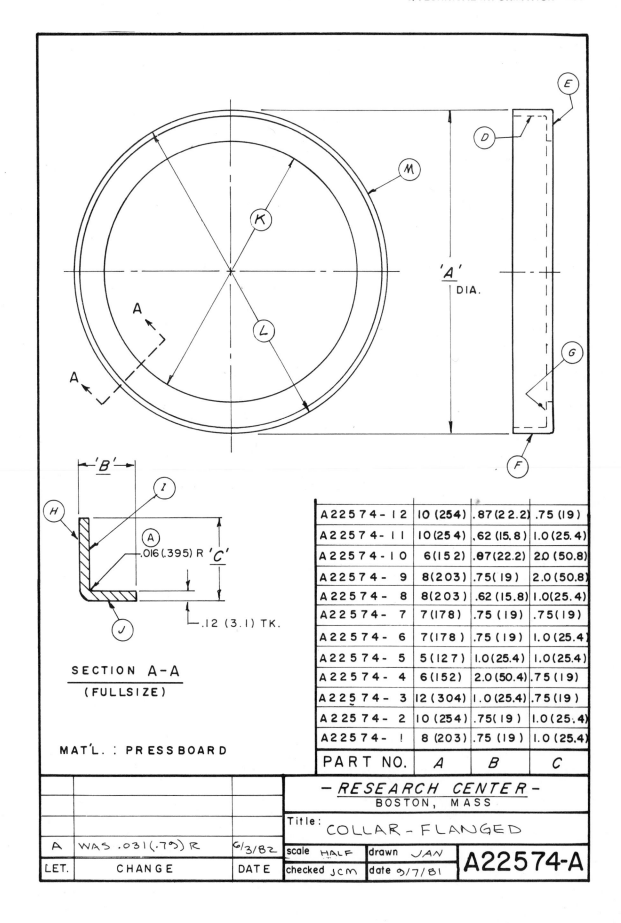

'A' DIA.

SECTION A-A
(FULLSIZE)

.016 (.395) R 'C'

.12 (3.1) TK.

MAT'L. : PRESSBOARD

PART NO.	A	B	C
A22574-12	10 (254)	.87 (22.2)	.75 (19)
A22574-11	10 (254)	.62 (15.8)	1.0 (25.4)
A22574-10	6 (152)	.87 (22.2)	2.0 (50.8)
A22574- 9	8 (203)	.75 (19)	2.0 (50.8)
A22574- 8	8 (203)	.62 (15.8)	1.0 (25.4)
A22574- 7	7 (178)	.75 (19)	.75 (19)
A22574- 6	7 (178)	.75 (19)	1.0 (25.4)
A22574- 5	5 (127)	1.0 (25.4)	1.0 (25.4)
A22574- 4	6 (152)	2.0 (50.4)	.75 (19)
A22574- 3	12 (304)	1.0 (25.4)	.75 (19)
A22574- 2	10 (254)	.75 (19)	1.0 (25.4)
A22574- 1	8 (203)	.75 (19)	1.0 (25.4)

- RESEARCH CENTER -
BOSTON, MASS

Title: COLLAR - FLANGED

A	WAS .031 (.79) R	6/3/82	scale HALF drawn JAN
LET.	CHANGE	DATE	checked JCM date 9/7/81

A22574-A

TABULAR DRAWINGS WORKSHEET

Using drawing A37116, answer the following questions. See end of chapter for answers.

1. What is the diameter of *'I'*? _____

2. What is the surface finish at surface *'J'*? _____

3. The 5/16-18 thread, (*K*) is it coarse or fine? _____

4. What does *3-B* in the 5/8-18-UN thread call-off mean? _____
   _____

5. What is dimension *'D'* for part A37116-2? _____
   _____

6. The 1/2-20 thread, (*E*) is it coarse or fine? _____

7. What manufacturing process is done on diameter *'F'*? _____

8. What manufacturing process is done on diameter *'G'*? _____

9. How many .12 (3.1) dia. holes are there? _____

10. How *wide* is the knurled surface? _____

11. How deep is the thread relief cut to? _____

12. What is the size of the largest undercut? _____

13. What is the surface finish of the 1.000 (25.400) diameter? _____
    _____

14. What is the overall length of part no. A37116-3? _____

15. What is dimension *'H'* of part no. A37116-1? _____

16. How deep are the four .12 (3.1) dia. holes? _____

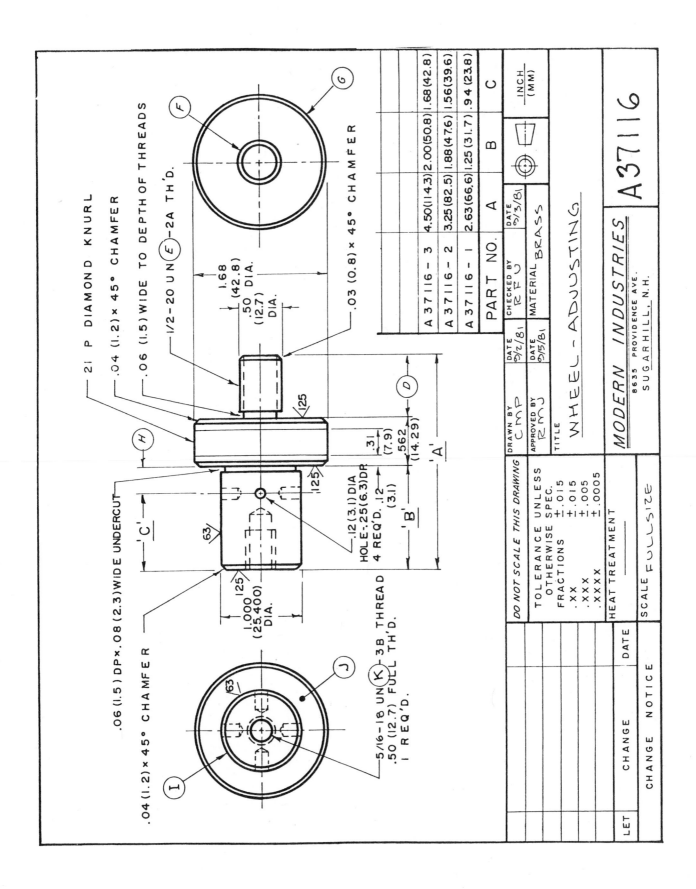

.06 (1.5) DP×.08 (2.3) WIDE UNDERCUT

.04 (1.2) × 45° CHAMFER

21 P DIAMOND KNURL

.04 (1.2) × 45° CHAMFER

.06 (1.5) WIDE TO DEPTH OF THREADS

1/2-20 UN Ⓔ -2A TH'D.

.03 (0.8) × 45° CHAMFER

1.68 (42.8)

.50 (12.7) DIA.

.125

.31 (7.9)

.562 (14.29)

.12 (3.1) DIA
HOLE-.25 (6.3) DR
4 REQ'D. .12 (3.1)

63

.125

1.000 (25.400) DIA.

5/16-18 UN Ⓚ -3B THREAD
.50 (12.7) FULL TH'D.
1 REQ'D.

63

PART NO.	A	B	C
A 37116 - 3	4.50 (114.3)	2.00 (50.8)	1.68 (42.8)
A 37116 - 2	3.25 (82.5)	1.88 (47.6)	1.56 (39.6)
A 37116 - 1	2.63 (66.6)	1.25 (31.7)	.94 (23.8)

INCH (MM)

DO NOT SCALE THIS DRAWING

TOLERANCE UNLESS
OTHERWISE SPEC.
FRACTIONS ±.015
.XX ±.015
.XXX ±.005
.XXXX ±.0005

HEAT TREATMENT

SCALE FULLSIZE

DRAWN BY CMP	DATE 6/2/81
APPROVED BY RMJ	DATE 6/5/81
CHECKED BY RFU	DATE 6/3/81
MATERIAL BRASS	

TITLE
WHEEL-ADJUSTING

MODERN INDUSTRIES
8635 PROVIDENCE AVE.
SUGARHILL, N.H.

A37116

| LET | CHANGE | DATE |
| CHANGE NOTICE | | |

DATUM PLANES

A datum is a reference point, line, surface, center or and important feature of an object, from which dimensions are derived. When datums are specified or implied, all features must be located from these datums. The illustration below uses the 'X', 'Y', and 'Z' datum planes. Where the datum planes are, is considered the "o" coordinate.

X, Y, AND Z COORDINATES

"X" measured all features from the *left* side, (as illustrated) to the *right*. "Y" measures all features from the *bottom* (as illustrated) to the *top*. "Z" measures all features from the *front* surface (as illustrated) to the rear surface. Think of: "X" as width, "Y" as height, and "Z" as depth.

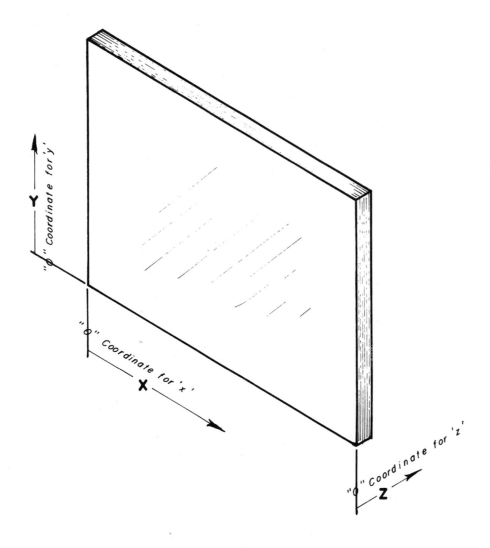

TABULAR DIMENSIONS

Study the tabular drawing below. Note the locations of the "o" coordinates for 'X', 'Y' and 'Z'. Each hole is lettered 'A', 'B', 'C', and 'D'. Each hole location is found by the coordinates 'X' and 'Y'. The depth of each hole is indicated by 'Z' coordinate.

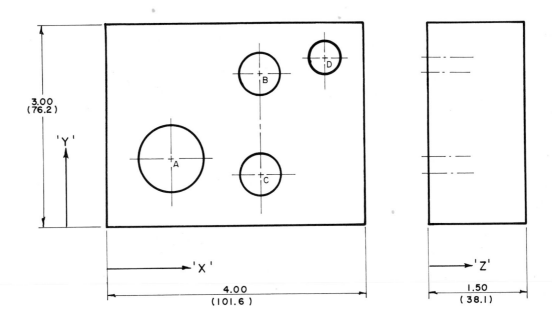

COORDINATE CHART

HOLE	DIA.	'X'	'Y'	'Z'
D	.500 (12.70)	3.344 (84.90)	2.500 (63.50)	1.250 (31.75)
C	.625 (15.88)	2.375 (60.30)	750 (19.05)	.750 (19.05)
B	.625 (15.88)	2.375 (60.30)	2.250 (57.10)	1.000 (25.40)
A	1.000 (25.40)	1.000 (25.40)	1.016 (25.79)	THRU

TABULAR DIMENSIONS WORKSHEET

Using plan no. A371192 answer the following questions. See end of chapter for answers.

1. Calculate dimension 'A'. _____

2. What is dimension 'B'? _____

3. What is the *maximum* limit allowed for dimension 'C'? _____

4. What is the *minimum* limit allowed for dimension 'D'? _____

5. Calculate dimension 'E'. _____

6. What is the *smallest* distance dimension 'F' can be and still be within tolerance?
_____

7. What is the radius at 'G'? _____

8. What is the *maximum* size dimension 'H' can be and still be within tolerance?
_____

9. How far apart are holes G1 and G1? _____
(Dimension 'I')? _____

10. What surface finish is surface 'J' to be machined to? _____

11. What is dimension 'K'? _____

12. What important point should be noted for holes E1 and E2? _____
_____

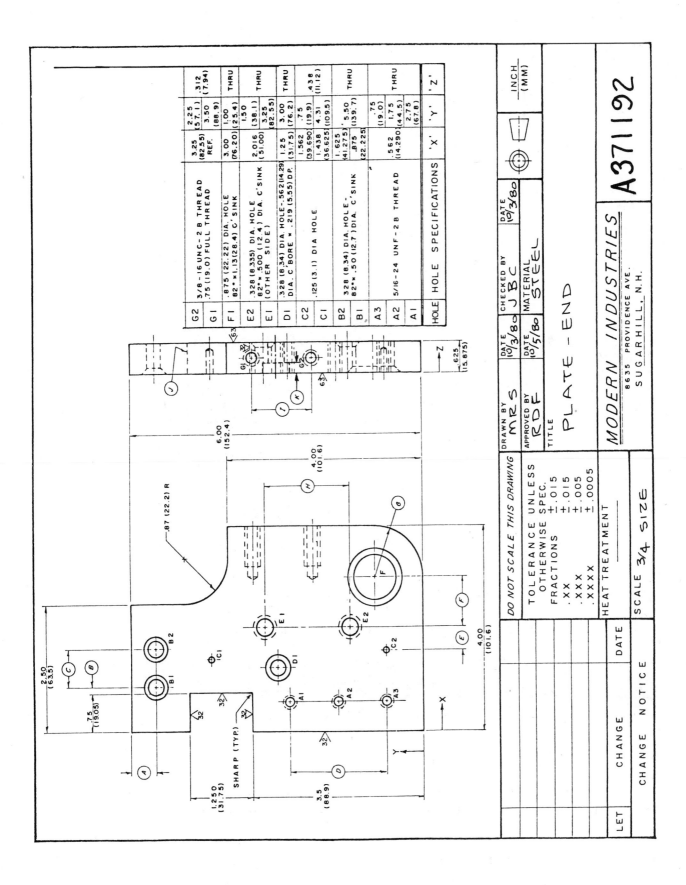

HOLE HOLE SPECIFICATIONS

HOLE		'X'	'Y'	'Z'
A1				
A2	5/16-24 UNF-2B THREAD	.562 (14.290)	2.75 (67.8)	THRU
A3			1.75 (44.5)	
B1	.328 (8.34) DIA. HOLE - 82°×.50 (12.7) DIA. C'SINK	.875 (22.225)	.75 (19.0)	
B2		1.625 (41.275)	5.50 (139.7)	THRU
C1	.125 (3.1) DIA HOLE	1.438 (36.625)	4.31 (109.5)	
C2		1.562 (39.690)	.75 (19.9)	
D1	.328 (8.34) DIA. HOLE-.562(14.29) DIA. C BORE × .219 (5.55) D.P.	1.25 (31.75)	3.00 (76.2)	THRU
E1	.328 (8.335) DIA. HOLE 82°×.500 (12.4) DIA. C'SINK (OTHER SIDE)	2.016 (51.00)	3.25 (82.55)	THRU
E2			1.50 (38.1)	
F1	.875 (22.22) DIA. HOLE 82°×1.13(28.4) C'SINK	3.00 (76.20)	1.00 (25.4)	THRU
G1	3/8-16UNC-2B THREAD .75 (19.0) FULL THREAD	3.25 (82.55) REF.	3.50 (88.9)	
G2		2.25 (57.1)	.312 (7.94)	

DO NOT SCALE THIS DRAWING

TOLERANCE UNLESS OTHERWISE SPEC.
FRACTIONS ± .015
.XX ± .015
.XXX ± .005
.XXXX ± .0005

HEAT TREATMENT

LET	CHANGE	DATE

CHANGE NOTICE

DRAWN BY MRS   DATE 10/3/80
CHECKED BY JBC   DATE 10/3/80
APPROVED BY RDF   DATE 10/5/80
MATERIAL STEEL

TITLE PLATE - END

MODERN INDUSTRIES
8635 PROVIDENCE AVE.
SUGARHILL, N.H.

INCH (MM)

A371192

SCALE 3/4 SIZE

GEOMETRIC TOLERANCING

Introduction

Up to this point, permissible variation of *size* has been used. Geometric tolerancing is the permissible variation of *shape*. Most shapes are broken down into basic geometric forms such as a plane (surface), cylinder, cone, square, hex, etc. As noted before, *nothing* can be made exactly the same *size* every time, and nothing can be made exactly the same *shape* every time, so a system of tolerancing is needed. For *size,* a system of limits and tolerances are incorporated; for shape, geometric tolerancing is used. Geometric tolerancing is used only where the shape could be critical to the function of the part, or could effect interchangeability of parts.

Some companies do not use geometric tolerancing, others use it extensively. The next few illustrations give a brief condensed idea of what geometric tolerancing is and basically how it works. Geometric tolerancing applies tolerancing to:

1. Flatness
2. Straightness
3. Cylindricity
4. Roundness
5. Parallelism
6. Perpendicularity
7. Angularity
8. Concentricity
9. Symmetry

Feature Control Note

Any and all feature control notes, must call-off what planes (views) the tolerances will be in respect too. Sometimes a tolerance is tied to more than one plane (view). Thus, the first, second, and auxiliary datum notes.

If any surface does not have a geometric tolerance specified, the form is allowed to vary within the given limits of size.

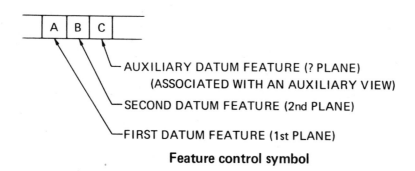

Feature control symbol

Flatness

This means that the *entire* surface must lie between two parallel planes that cannot be more than the specified tolerance apart. The symbol for flatness is a parallelogram.

(Note: Flatness is used on flat surfaces only).

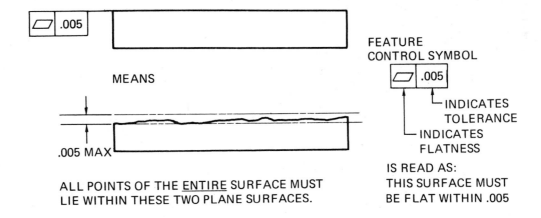

MEANS

.005 MAX

ALL POINTS OF THE <u>ENTIRE</u> SURFACE MUST
LIE WITHIN THESE TWO PLANE SURFACES.

FEATURE
CONTROL SYMBOL

☐ .005

└ INDICATES
TOLERANCE
└ INDICATES
FLATNESS

IS READ AS:
THIS SURFACE MUST
BE FLAT WITHIN .005

Straightness

This means that the *entire* surface must be straight within given limit. (Note straightness is used on cylinder or cone surfaces only).

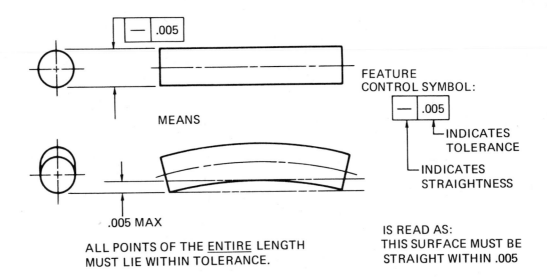

MEANS

.005 MAX

ALL POINTS OF THE <u>ENTIRE</u> LENGTH
MUST LIE WITHIN TOLERANCE.

FEATURE
CONTROL SYMBOL:

— .005

└ INDICATES
TOLERANCE
└ INDICATES
STRAIGHTNESS

IS READ AS:
THIS SURFACE MUST BE
STRAIGHT WITHIN .005

Cylindricity

This means that the *entire* length of the cylinder must lie between two parallel planes that cannot be more than a specified tolerance apart. (Note: Both circles must be about the same axis).

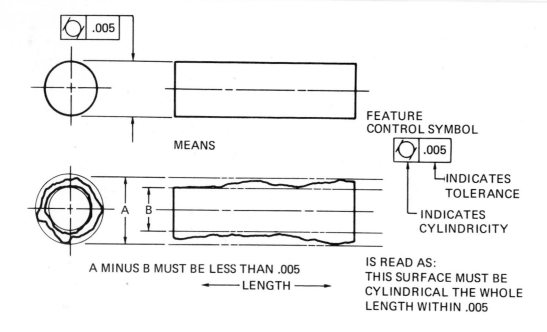

Geometric tolerancing, cylindricity

Roundness

This means that every point anywhere along the entire length of the cylinder must be round within specified tolerance (on dia.). The difference between cylindricity and roundness is the former involves overall length and the latter involves the object being round the entire length. (Note: Most round object would be manufactured on a lathe and thus would revolve around the center line or "reference" axis).

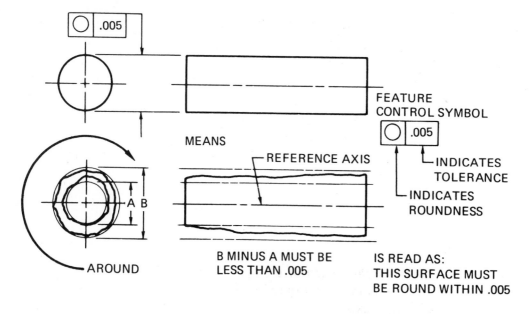

Geometric tolerancing, roundness

Parallelism

This means that an entire surface must lie between two parallel planes that cannot be more than specified tolerance apart, parallel to, and in relation to a given surface or datum. (Note: The bottom surface is datum "A"). Remember a datum could be a surface, point, centerline or any feature on the object.

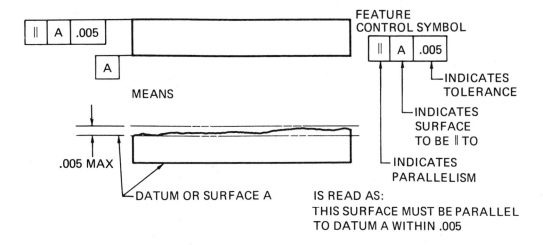

Perpendicularity

This means that an *entire* surface must lie between two parallel planes that cannot be more than specified tolerance apart. They are perpendicular to a given surface or datum. (Note: The upright surface is datum "A").

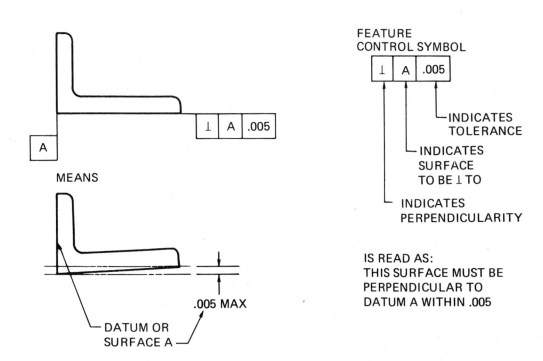

If any surface does not have a geometric tolerance specified, the form is allowed to vary within the given limits of size.

Angularity

This means that the *entire* surface must lie between two parallel planes that are at the *true* angle in relation to a specified surface or datum. (Note: More than one datum can be used) with specified tolerance.

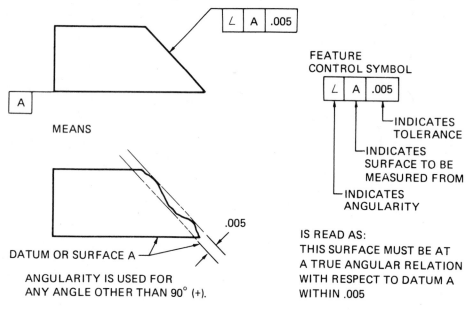

MEANS

DATUM OR SURFACE A

ANGULARITY IS USED FOR
ANY ANGLE OTHER THAN 90° (+).

FEATURE
CONTROL SYMBOL

∠ A .005

└ INDICATES
TOLERANCE

└ INDICATES
SURFACE TO BE
MEASURED FROM

└ INDICATES
ANGULARITY

IS READ AS:
THIS SURFACE MUST BE AT
A TRUE ANGULAR RELATION
WITH RESPECT TO DATUM A
WITHIN .005

Concentricity

This means that the axis of one feature or diameter must lie within a cylindrical tolerance zone (which is concentric) to the datum axis of another feature or diameter, with a specified tolerance.

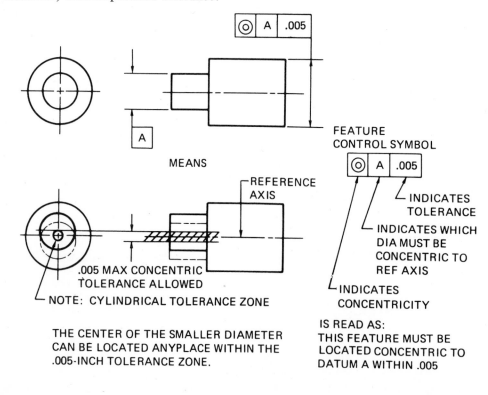

MEANS

REFERENCE
AXIS

.005 MAX CONCENTRIC
TOLERANCE ALLOWED

NOTE: CYLINDRICAL TOLERANCE ZONE

THE CENTER OF THE SMALLER DIAMETER
CAN BE LOCATED ANYPLACE WITHIN THE
.005-INCH TOLERANCE ZONE.

FEATURE
CONTROL SYMBOL

⊚ A .005

└ INDICATES
TOLERANCE

└ INDICATES WHICH
DIA MUST BE
CONCENTRIC TO
REF AXIS

└ INDICATES
CONCENTRICITY

IS READ AS:
THIS FEATURE MUST BE
LOCATED CONCENTRIC TO
DATUM A WITHIN .005

Symmetry

This means that the entire feature must lie between two parallel planes that cannot be more than specified tolerance apart, symmetrically, and in regard to a surface(s) or datum. The specified tolerance in this case must be *equally spaced between the datum,* or surface(s).

If any surface(s) does not have a geometric tolerance specified, the form is allowed to vary within the given limits of size.

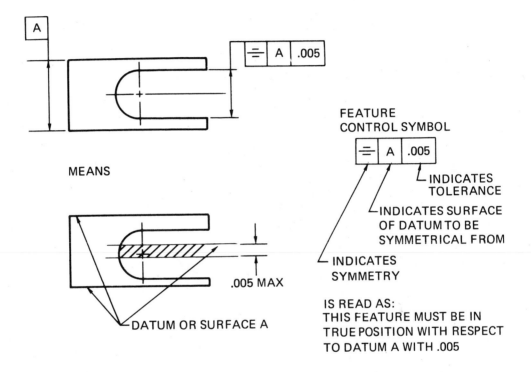

MEANS

.005 MAX

DATUM OR SURFACE A

FEATURE
CONTROL SYMBOL

INDICATES
TOLERANCE

INDICATES SURFACE
OF DATUM TO BE
SYMMETRICAL FROM

INDICATES
SYMMETRY

IS READ AS:
THIS FEATURE MUST BE IN
TRUE POSITION WITH RESPECT
TO DATUM A WITH .005

MODIFIERS

Sometimes a modifier is added to the control note:

Ⓢ Means: "Regardless of feature size" (RFS)

This reads as: Parallel to surface "A", regardless of feature size and within .005 inch.

Ⓜ Means: "Maximum material condition" (MMC)

This reads as: Parallel to surface "A" when it is at maximum material condition only and within a .005 inch.

The note T.I.R. stands for "total indicator reading." This means the part to be checked must be round (the full 360°) to within .005 T.I.R. The part is set up so it can rotate about a fixed center line. An indicator is mounted above it and set on "0". The part is then rotated and the indicator must *not* move than .005 total *both* directions. (If it does, it is *not* within tolerance.)

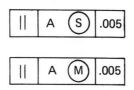

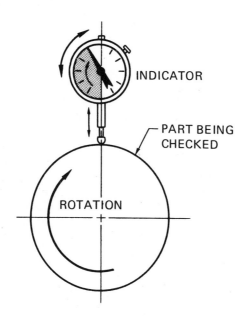

GEOMETRIC TOLERANCING WORKSHEET

Carefully study each *control note* on this sample drawing (1-8) and, in the space provided, explain *in full* what each control note means. (Note, this example has more control notes than usually found on a simple drawing.) See end of chapter for answers.

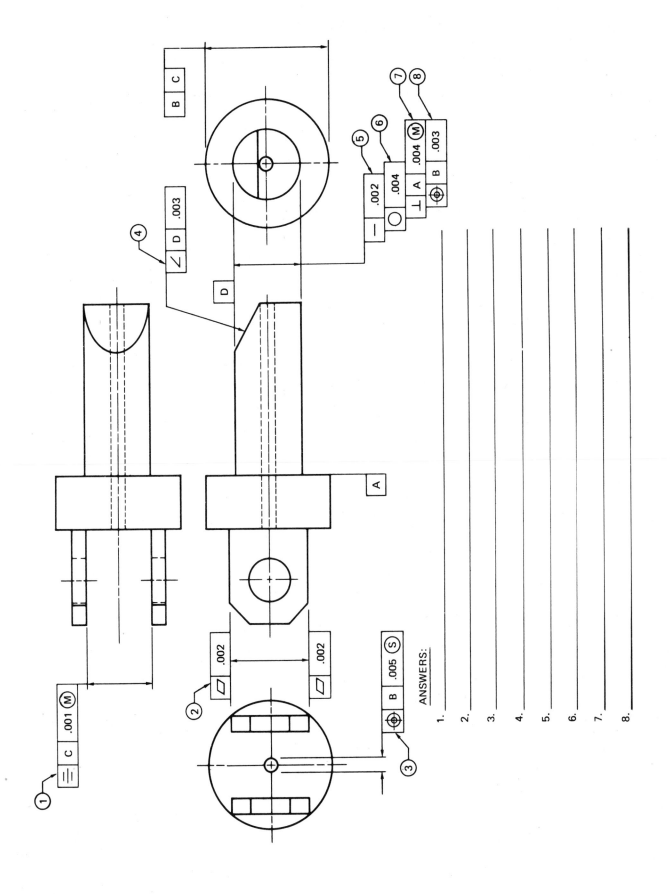

ANSWERS:

1. _____

2. _____

3. _____

4. _____

5. _____

6. _____

7. _____

8. _____

TRUE POSITIONING

Using the *normal* bilateral system of dimensioning a hole would be dimensioned as illustrated with limits of + or − .015. The draftsperson really wants the center of the hole 1.00 up & over from the edges *but* will except + or −.015 limits.

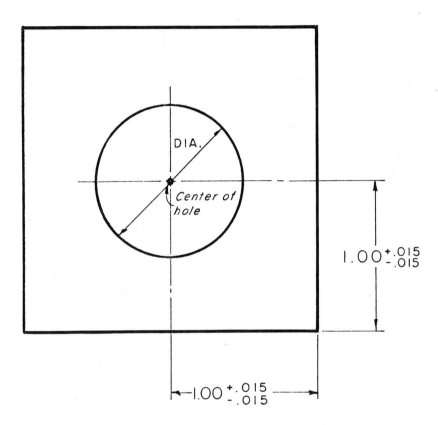

The *tolerance zone* would form a square area that, in effect, establishes boundaries that the center of the circle must *not* go outside of.

Note the .030 tolerance zone extends the *full depth* of the hole.

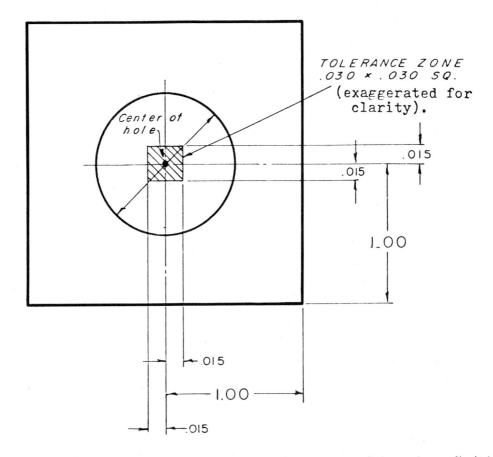

Example 1. These four examples illustrate the worst conditions, (max. limits). Vertically and horizontally, the center of the hole can be from the design size and still be within limits.

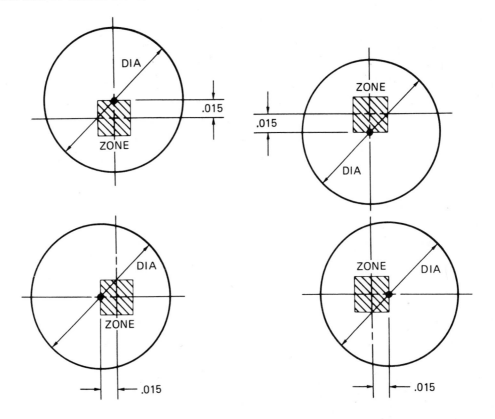

Example 2. These four examples illustrate the worst conditions, (max. limits). Diagonal across corners, the center of the old can be from the design size and still be within limits.

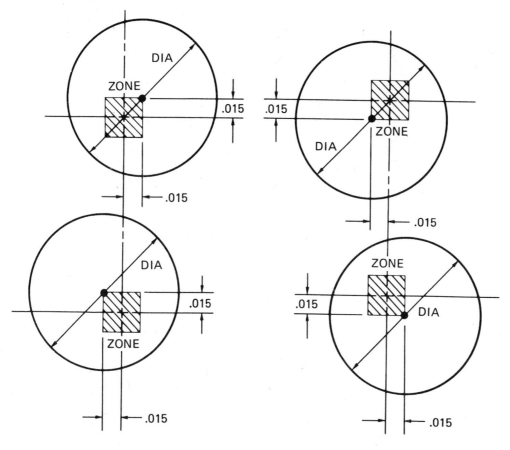

In Example 1, the maximum distance from the design size is .015 and still be within limits:

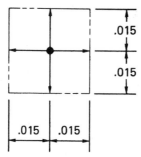

In Example 2, the maximum distance from the design size is much more:

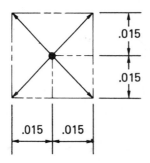

Using simple trigonometry, it is possible to calculate the exact diagonal distance from the center of the hole, or .021 radii. In effect, there are *two* sizes or limits used in locating the center of the hole. +/– .015 one way and +/– .012 the other way, using bilateral tolerancing.

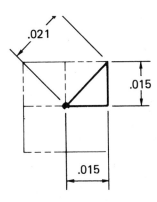

If .021 is within tolerance, why not allow .021 radii *all around.* (A circle has 57% more area than a square.) This will reduce scrap, reduce inspection time, and reduce cost, and, in effect, still allow the *exact* same limits as the bilateral system did (across corners).

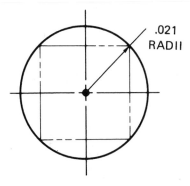

The symbol for T.P. (true position is $\bigoplus$)

When comparing the bilateral tolerancing system with the true positional (T.P.) system, it is easy to see the tolerance zone using T.P. is much larger, meaning fewer rejections of parts and lower cost.

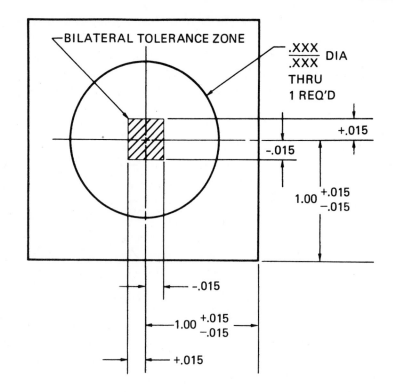

There is a 57% larger tolerance zone using T.P.

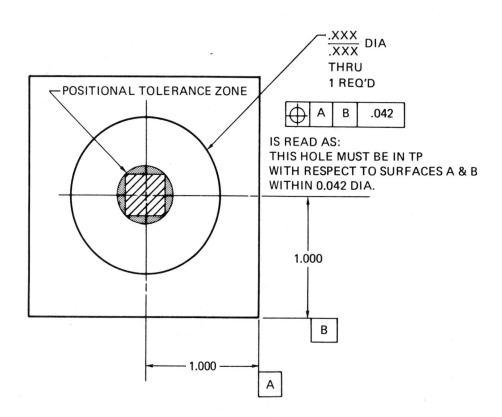

This is a list of most of the standard symbols used in engineering. Be sure you know and understand what each symbol means.

SYMBOLS

▱	FLATNESS
—	STRAIGHTNESS
⌭	CYLINDRICITY
○	ROUNDNESS
‖	PARALLELISM
⊥	PERPENDICULARITY
∠	ANGULARITY
◎	CONCENTRICITY
⩦	SYMMETRY
Ⓢ	REGARDLESS OF FEATURE SIZE
Ⓜ	MAXIMUM MATERIAL CONDITION
⌖	TRUE POSITION

T.P. reduces scrap, reduces inspection time, simplifies calculations, and most important reduces cost.

As with geometric tolerancing, many companies do not use true positioning, thus not too much detail will be given to it at this time. In the event you work for a company that does use it, this brief outline will give you a *basic* working knowledge of T.P. There are many good books available that explain *all* areas of T.P.

1st ANGLE PROJECTION

1st. angle projection is used by all countries *except* the U.S., Canada, and England. It is similar to 3rd angle projection except the views are rotated in an opposite direction from the front view. Study comparison illustrations below. Both start from the *front* view.

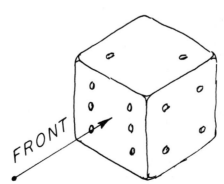

The Title Block indicates which angle of projection is used. Most countries using 1st angle projection also use the metric system of measurement as the design dimensions, thus all tolerances must be given using the metric system.

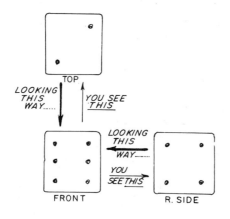

3rd Angle projection

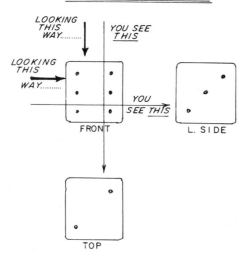

1st Angle projection

TITLE BLOCK SYMBOL

Up to this time, only *3rd angle projection* has been illustrated and used. Some companies use only one of the systems, others use *both* systems—thus it is important the craftsperson know and understand *both* systems.

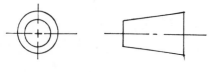

3rd Angle symbol

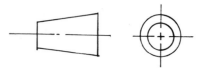

1 st Angle symbol

THREE VIEW DRAWING (1st Angle Projection)

Compare this sketch to the one describing 3rd angle projection.

T
O
P

V
I
E
W

FRONT VIEW →

← RIGHT SIDE VIEW

Pictorial View

A three-view, 1st angle projection drawing is similar to a three-view, 3rd angle projection, *except* the views are rotated in different directions. Starting with the *front view,* the object is rotated to the *right,* which shows the *left side.* Again, from the front view rotate the object *downward,* which will show the *top view.*

FRONT VIEW..............

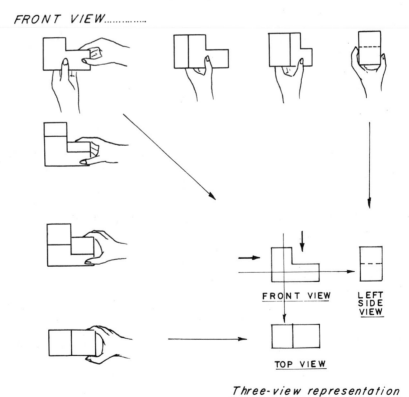

FRONT VIEW

LEFT
SIDE
VIEW

TOP VIEW

Three-view representation

Note: The *top* view is directly *below* the front view and the *left* side view is directly to the *right* of the front view.

1st ANGLE PROJECTION WORKSHEET I

Using the *1st angle projection,* three-view drawing on the opposite page, answer the following questions. (Answer all questions using millimeters). See end of chapter for answers.

1. What is dimension 'A'? _____

   _____

2. Calculate dimension 'B'. _____

   _____

3. How many inches is dimension 'C'? _____

   _____

4. Assuming the 22.2 (7/8) notch is in the center of the part, what would dimension 'D' be? _____

   _____

5. Surface 'E' in view 1 is what surface in view 3? _____

   _____

6. Surface 'F' in view 3 is what surface in view 1? _____

   _____

7. Surface 'G' in view 3 is what surface in view 1? _____

   _____

8. Surface 'H' in view 2 is what surface in view 1? _____

   _____

9. Surface 'I' in view 3 is what surface in view 1? _____

   _____

10. What is view 1 called? _____

    _____

11. What is view 2 called? _____

    _____

12. What is view 3 called? _____

    _____

Note: Compare the drawing on the opposite page with the 3rd angle projection drawing of the same object in the Three-view Drawing Worksheet I in Chapter 4.

1st ANGLE PROJECTION

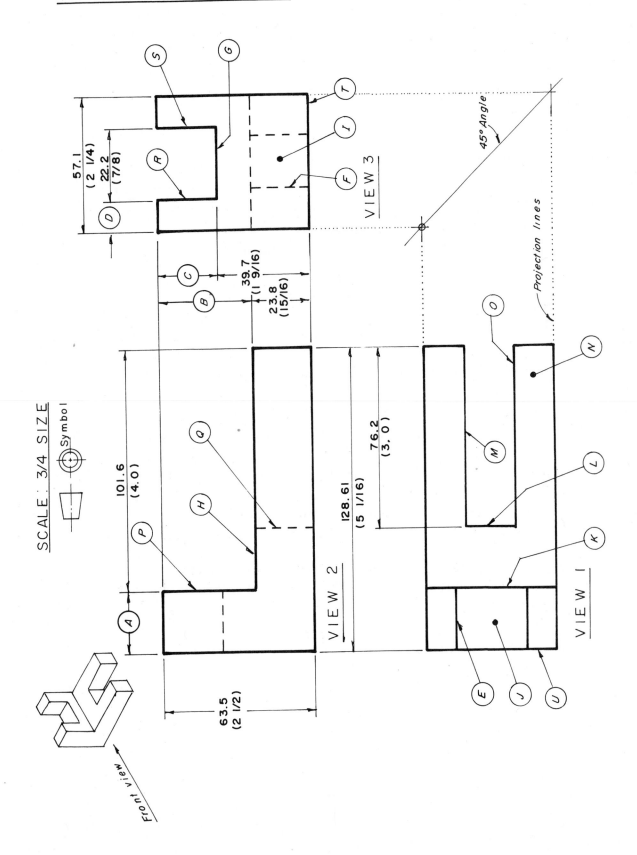

SCALE : 3/4 SIZE

Symbol

VIEW 1

VIEW 2

VIEW 3

Front view

45° Angle

Projection lines

57.1 (2 1/4)
22.2 (7/8)
39.7 (1 9/16)
23.8 (15/16)
101.6 (4.0)
128.61 (5 1/16)
76.2 (3.0)
63.5 (2 1/2)

1st ANGLE PROJECTION WORKSHEET II

Using the three-view drawing on the opposite page answer the following questions. See end of chapter for answers.

1. Calculate dimension 'A'. _____

2. What is dimension 'B'? _____

3. What is the length, height, and depth of this object? _____

4. Dimension 'C' is equal to? _____

5. What is dimension 'D'? _____

6. Calculate dimension 'E'. _____

7. What is dimension 'F'? _____

8. What is dimension 'G'? _____

9. Dimension 'H' is equal to? _____

10. Line 'L' in the top view is what surface in the front view? _____

11. Line 'Q' in the front view is what surface in the top view? _____

12. Line 'R' in the front view is what surface in the left side view? _____
    _____

13. Line 'Z' in the left side view is what surface in the top view? _____
    _____

14. Surface 'W' in the left side view is what surface in the front view? _____
    _____

15. Surface 'W' in the left side is what surface in the top view? _____

16. Surface 'I' in the top view represents what surface in the left side view? ___
    _____

17. Line 'U' in the left side view represents what surface in the top view? _____
    _____

18. In the top view, a radii is noted by a 'R'. What is this radius? _____
    _____

Note: Compare the drawing on the opposite page with the 3rd angle projection drawing of the same object in the three-view Drawing Worksheet II in chapter 4.

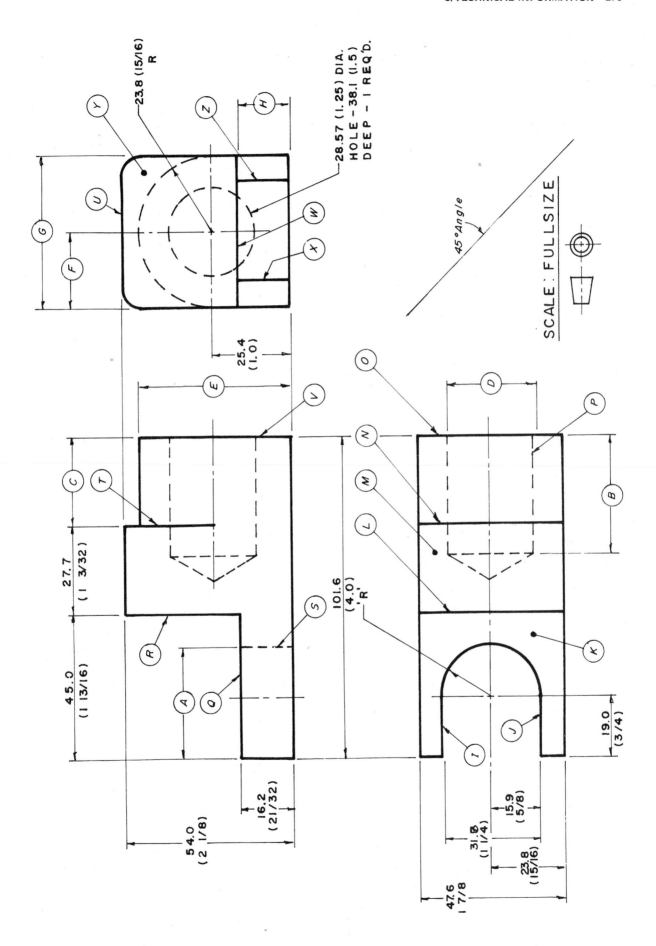

23.8 (15/16) R

28.57 (1.25) DIA.
HOLE – 38.1 (1.5)
DEEP – I REQ'D.

45° Angle

SCALE : FULLSIZE

25.4 (1.0)

27.7 (1 3/32)

45.0 (1 13/16)

101.6 (4.0)

R

16.2 (21/32)

54.0 (2 1/8)

19.0 (3/4)

15.9 (5/8)

31.8 (1 1/4)

23.8 (15/16)

47.6 1 7/8

EVALUATION

Explain in full detail the meanings of the feature control notes below. See end of chapter for answers.

1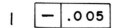

_____

2

_____

3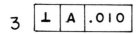

_____

4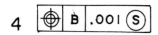

_____

5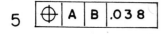

_____

6 T. I. R.

_____

7. Briefly explain what the 'X', 'Y', and 'Z' coordinates are. _____
   _____

8. Exactly what is geometric tolerancing? _____
   _____

9. What is a feature control note? _____
   _____

10. Describe a tabular drawing. _____
    _____

11. What does the modifier "S" indicate? _____
    _____

12. How much *larger* is the tolerance zone using T.P. compared to the conventional coordinate dimensioning system? _____
    _____

Using drawing A-374593, answer the following questions.

1. What angle of projection is used on this drawing? _____
    _____

2. What system of measuring does the controlling dimension symbol indicate?
    _____

3. What is dimension 'A'? _____

4. Surface 'C' in the top view is what surface in the left side view? _____
    _____

5. Calculate radius 'D'. _____

6. Surface 'G' in the top view is what surface in the left side view? _____
    _____

7. Surface 'F' in the top view is what surface in the left side view? _____
    _____

8. Calculate dimension 'H'. _____

9. Surface 'I' in the front view is what surface in the left side view? _____
    _____

10. Surface 'K' in the front view is what surface in the top view? _____
    _____

11. Hole 'a' is represented in the left side view by what letter? _____
    _____

12. What is dimension 'M'? _____

13. What is dimension 'O'? _____

14. Calculate dimension 'P'. _____

15. 'R' is the center line of what? _____

16. What does the wavy line at 'S' indicate? _____

17. What is the radii at 'Z'? _____

18. The hidden line at 'N' represents what? _____

19. What is the center to center holes distance? _____
    _____

20. What is the largest size the 15.8 (5/8) dia. hole can be made and pass inspection? _____

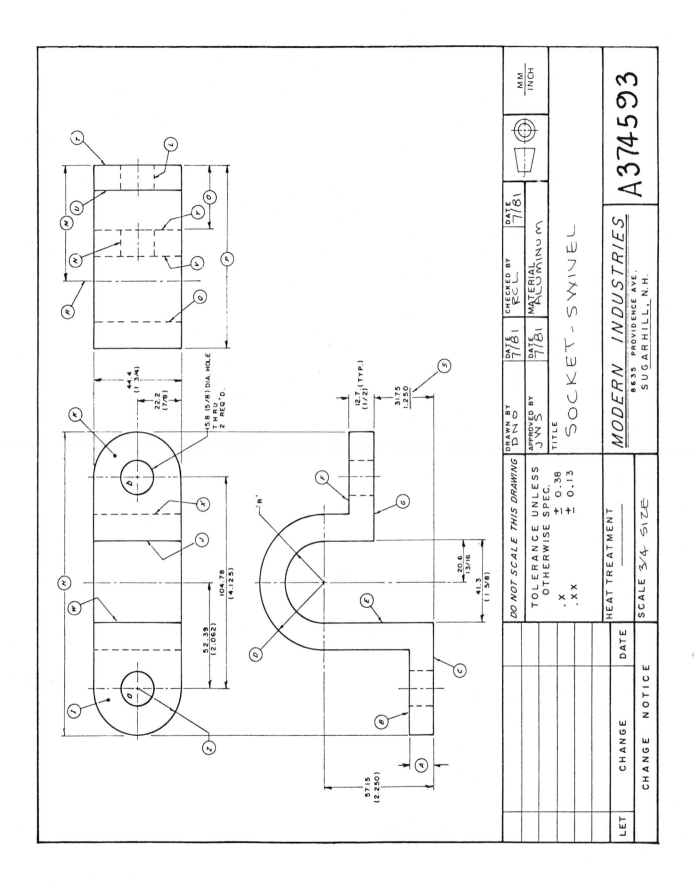

MM / INCH

DRAWN BY DNG   DATE 7/81
APPROVED BY JWS   DATE 7/81
CHECKED BY RCL   DATE 7/81
MATERIAL ALUMINUM

TITLE
SOCKET-SWIVEL

MODERN INDUSTRIES
8635 PROVIDENCE AVE.
SUGARHILL, N.H.

A374593

DO NOT SCALE THIS DRAWING

TOLERANCE UNLESS
OTHERWISE SPEC.
.X   ± 0.38
.XX   ± 0.13

HEAT TREATMENT

SCALE 3/4 SIZE

LET   CHANGE   DATE

CHANGE NOTICE

44.4
(1 3/4)

22.2
(7/8)

15.8 (5/8) DIA. HOLE
THRU
2 REQ'D.

104.78
(4.125)

52.39
(2.062)

57.15
(2.250)

12.7 (TYP.)
(1/2)

31.75
1.250

20.6
13/16

41.3
(1 5/8)

ANSWERS TO TERMS TO KNOW AND UNDERSTAND

Tabular Drawings—A drawing that uses 'reference letters' in place of variable dimensions. A table lists these reference letters with corresponding dimensions.

Datum Planes—A point, line, surface center, or an important feature from which dimensions are derived.

"X"—Measurements from left to right (width).

"Y"—Measurements from bottom to top (height).

"Z"—Measurements from front to back (depth).

Tabular Dimensions—Dimensions taken from an established datum.

Geometric Tolerance—Permissable variation of shape.

Feature Control Note—A symbol, or "call-off" used to give tolerances to shapes and surfaces.

T.I.R.—Total indicator reading.

Ⓢ-Modifier, *"regardless of feature size."*

Ⓜ-Modifier, *"maximum material condition."*

True Positioning—A method of dimensioning whereby the tolerance zone is a circle (57% more area than a square zone).

1st Angle Projection—Similar to 3rd angle projection except views are rotated in different positions—used by all countries except U.S., Canada, and England.

TABULAR DRAWINGS WORKSHEET ANSWERS

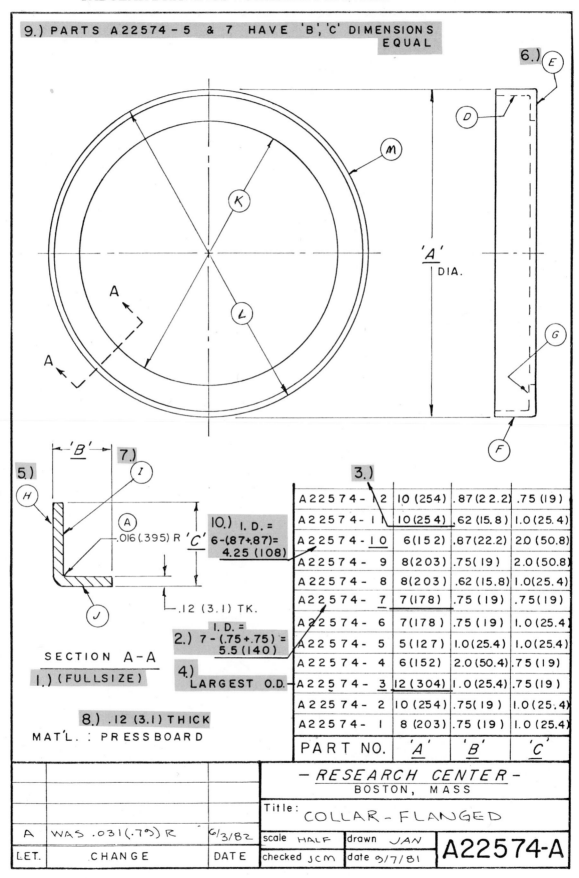

9.) PARTS A22574-5 & 7 HAVE 'B','C' DIMENSIONS EQUAL

6.)

'A' DIA.

5.)

'B'   7.)

10.) I. D. =
6-(.87+.87)=
4.25 (108)

.016 (.395) R 'C'

.12 (3.1) TK.

2.) I. D. =
7 - (.75+.75) =
5.5 (140)

SECTION A-A

1.) (FULLSIZE)

4.) LARGEST O.D.

8.) .12 (3.1) THICK

MAT'L. : PRESSBOARD

3.)

PART NO.	'A'	'B'	'C'
A22574-12	10 (254)	.87 (22.2)	.75 (19)
A22574-11	10 (254)	.62 (15.8)	1.0 (25.4)
A22574-10	6 (152)	.87 (22.2)	2.0 (50.8)
A22574-9	8 (203)	.75 (19)	2.0 (50.8)
A22574-8	8 (203)	.62 (15.8)	1.0 (25.4)
A22574-7	7 (178)	.75 (19)	.75 (19)
A22574-6	7 (178)	.75 (19)	1.0 (25.4)
A22574-5	5 (127)	1.0 (25.4)	1.0 (25.4)
A22574-4	6 (152)	2.0 (50.4)	.75 (19)
A22574-3	12 (304)	1.0 (25.4)	.75 (19)
A22574-2	10 (254)	.75 (19)	1.0 (25.4)
A22574-1	8 (203)	.75 (19)	1.0 (25.4)

— RESEARCH CENTER —
BOSTON, MASS

Title: COLLAR - FLANGED

LET.	CHANGE	DATE
A	WAS .031 (.79) R	6/3/82

scale HALF	drawn JAN
checked JCM	date 9/7/81

A22574-A

TABULAR DRAWINGS WORKSHEET ANSWERS

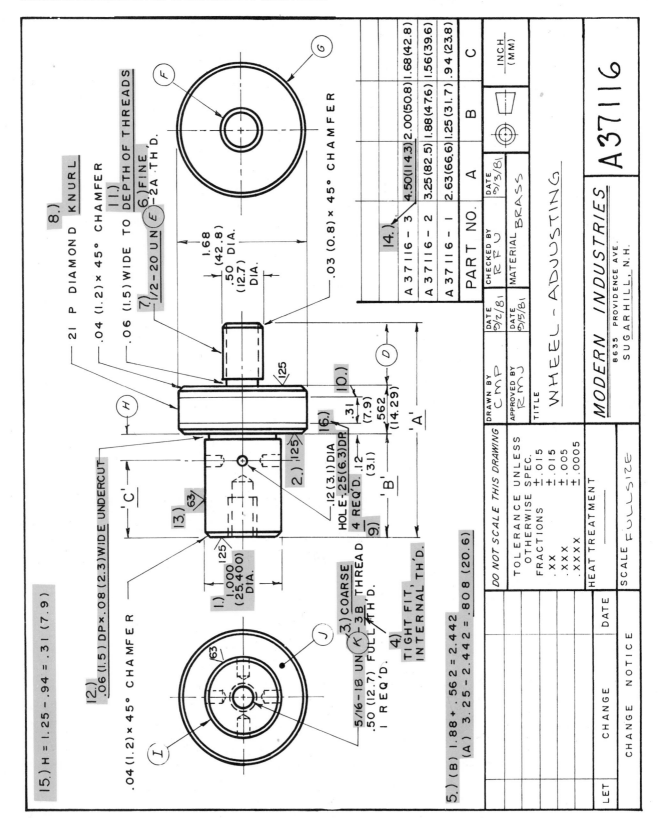

TABULAR DIMENSIONS WORKSHEET ANSWERS

Hole Specifications Table

Hole	HOLE SPECIFICATIONS	'X'	'Y'	'Z'
G2	3/8-16 UNC-2B THREAD .75(19.0) FULL THREAD	3.25(82.55) REF.	2.25(57.1)	.312(7.94)
G1	.875(22.22) DIA. HOLE 82°×1.13(28.4) C'SINK	3.00(76.20)	3.50(88.9)	THRU
F1			1.00(25.4)	THRU
E2	.328(8.335) DIA HOLE 82°×.500(12.4) DIA. C'SINK (OTHER SIDE) 12.)	2.016(51.00)	1.50(38.1)	THRU
E1	.328(8.34) DIA. HOLE-.562(14.29) DIA. C'BORE × .219(5.55) D.P. 11.)	1.25(31.75)	3.00(76.2)	THRU
D1	.328(8.34) DIA HOLE 82°×.50(12.7) DIA. C'SINK	1.562(39.690)	.75(19.9)	THRU
C2	.125(3.1) DIA HOLE	1.438(36.625)	4.31(109.5)	.438(11.12)
C1		1.625(41.275)	5.50(139.7)	THRU
B2	.328(8.34) DIA HOLE 82°×.50(12.7) DIA. C'SINK	.875(22.225)	1.75(44.5)	THRU
B1			.75(19.0)	THRU
A3	5/16-24 UNF-2B THREAD	.562(14.290)	2.75(67.8)	THRU
A2				
A1				

Worksheet Answers

1.) 6.00 − 5.50 = .50 (12.7)

2.) .875 − .75 = .125 (3.1)

3.) 1.625 − .875 = .750 + .005 = .755 (19.177)

4.) 2.75 − .75 = 2.00 − .015 = 1.985 (50.419)

5.) 2.016 − 1.562 = .454 (11.510)

6.) 3.0 − .015 = 2.985 MIN. 2.016 + .005 = 2.021 MAX. 2.985 − 2.021 = .964 (24.486)

7.) 4.00 − 3.0□ = 1.00 (25.4) R

8.) 3.25 + .015 = 3.265 MAX. 1.50 − .015 = 1.485 MIN. 1.780 (45.212)

9.) 3.50 − 2.25 = 1.25 (31.75)

10.) 32 SHARP (TYP.)

11.)

12.)

Title Block

MODERN INDUSTRIES
8635 PROVIDENCE AVE.
SUGARHILL, N.H.

A371192

INCH (MM)

TITLE: PLATE — END

DRAWN BY MRS — DATE 10/3/80
APPROVED BY RDF — DATE 10/5/80
CHECKED BY JBC — DATE 10/3/80
MATERIAL: STEEL

HEAT TREATMENT

SCALE 3/4 SIZE

DO NOT SCALE THIS DRAWING

TOLERANCE UNLESS OTHERWISE SPEC.
FRACTIONS ±.015
.XX ±.015
.XXX ±.005
.XXXX ±.0005

CHANGE NOTICE

LET	CHANGE	DATE

Additional dimensions: 6.00 (152.4), 4.00 (101.6), .87 (22.2) R, .625 (15.875), 2.50 (63.5), .75 (19.05), 1.250 (31.75), 3.5 (88.9), .63, 32

GEOMETRIC TOLERANCING WORKSHEET ANSWERS

1. The two 'ears' must be in true position with respect to surface "C" and within .001 when they are at *maximum material condition* (Note the *"M"* modifier).
2. This surface must be flat the whole length within .002.
3. The hole must be located concentric to surface "B" within .005 *regardless of size* (Note the "S" modifier).
4. This angle must be at the true specified angle with respect to surface "D" and within .003.
5. This diameter must be straight within its full length and within .002.
6. This diameter must be round the entire length and within .004.
7. This diameter must be perpendicular to surface "A" and within .004 at *maximum material condition* (Note the *"M"* modifier).
8. This diameter must be concentric to surface "B" and within .003.

1st ANGLE PROJECTION WORKSHEET I ANSWERS

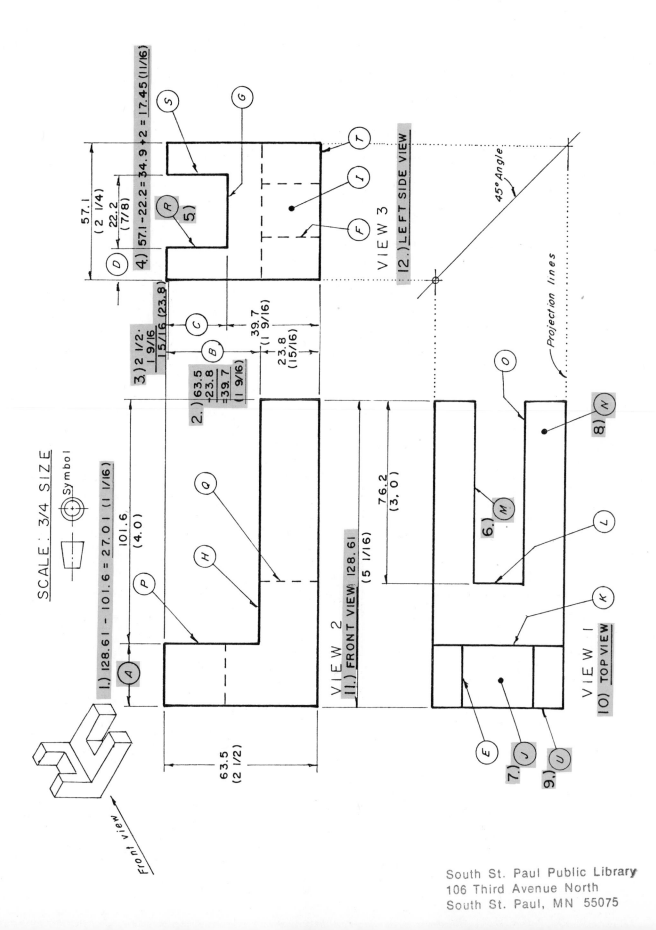

SCALE : 3/4 SIZE

Symbol

1.) 128.61 – 101.6 = 27.01 (1 1/16)   A

2.) 63.5 – 23.8 = 39.7 (1 9/16)

3.) 2 1/2 · 1 9/16 · 5/16 (23.8)

4.) 57.1 – 22.2 = 34.9 ÷ 2 = 17.45 (11/16)

5.) R

6.) M

7.) J

8.) N

9.) U

10.) TOP VIEW

VIEW 1

11.) FRONT VIEW 128.61

VIEW 2

12.) LEFT SIDE VIEW

VIEW 3

45° Angle

Projection lines

Front view

57.1 (2 1/4)

22.2 (7/8)

39.7 (1 9/16)

23.8 (15/16)

101.6 (4.0)

76.2 (3.0)

63.5 (2 1/2)

A B C D E F G H I J K L M N O P Q R S T U

South St. Paul Public Library
106 Third Avenue North
South St. Paul, MN 55075

1st ANGLE PROJECTION WORKSHEET II ANSWERS

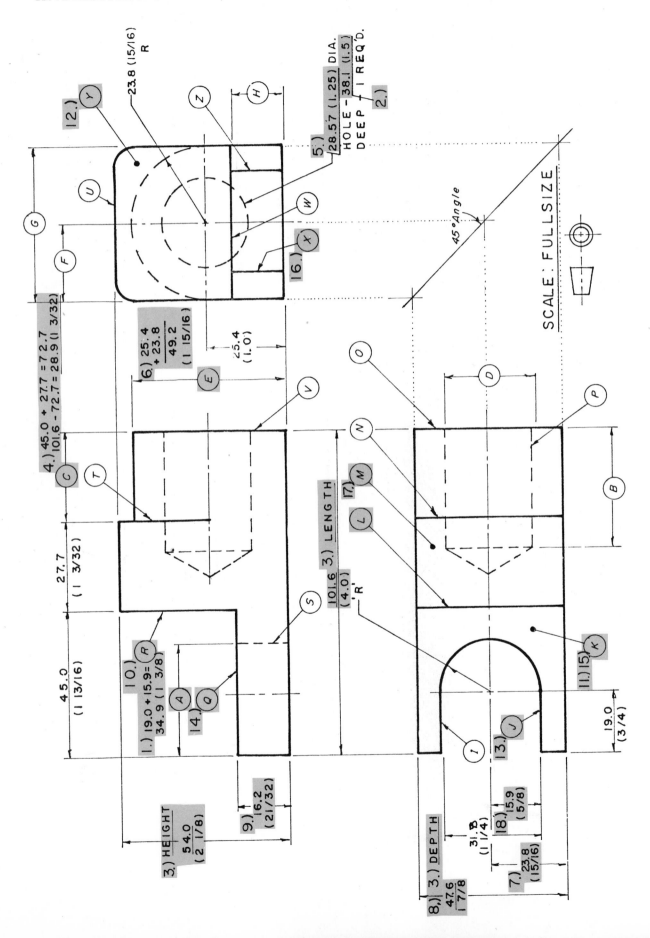

EVALUATION ANSWERS

1. This surface must be *straight* within .005 inch.
2. This surface must be *parallel* to surface 'A' when it is at maximum material condition only and must be within .002 inch.
3. This surface must be perpendicular to surface 'A' within .010 inch.
4. This hole must be located *concentric* to surface 'B' within .001 inch, regardless of size.
5. This hole must be in true position with respect to surface 'A' and 'B' within .038 dia.
6. Total indicator reading.
7. 'X' indicates measurements from left to right (width) 'Y' indicated measurements from bottom to top (height) 'Z' indicated measurements from front to back (depth) from noted datum planes.
8. Geometric tolerancing is the permissible variation of *shape.*
9. A feature control note is a symbol, "call-off" used to give exact tolerances with respect to shapes and surfaces.
10. A tabular drawing is a drawing that uses reference letters in place of variable dimensions. A table lists these reference letters with corresponding dimensions.
11. The modifier *"S"* means regardless of feature size.
12. 57%

EVALUATION ANSWERS

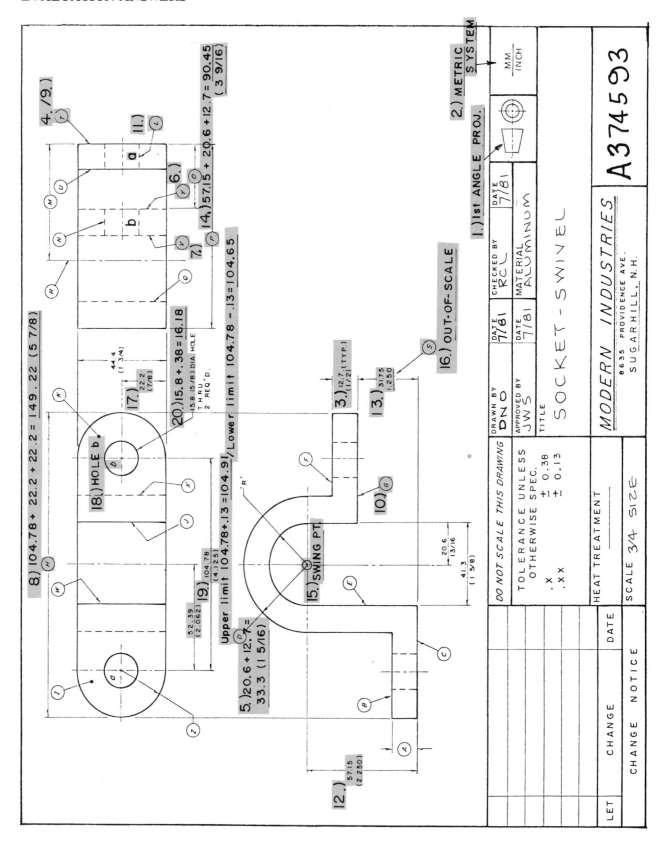

UNIT 9

PRECISION MEASUREMENT

Objective: To learn to use and read micrometers and vernier calipers.

TERMS TO KNOW AND UNDERSTAND

See end of chapter for answers.

Pocket Steel Ruler _____

English System _____

Metric System _____

Meter _____

Kilometer _____

Millimeter _____

Calipers _____

Micrometer _____

Spindle _____

Sleeve _____

Thimble _____

Frame _____

Anvil _____

Telescoping Gauges _____

Depth Gauges _____

Vernier Calipers _____

Vernier _____

Inside/outside Measurements (calipers) _____

MEASURING

This unit illustrates how to make precise measurements using both the inch (English system) and the millimeter (metric system). The metric system uses the *meter* (m) as its basic dimension. It is 3.281 feet long or about 3 3/8 inches longer than a yardstick. Its multiples, or parts, are expressed by adding prefixes. These prefixes represent equal steps of 1000 parts. The prefix for a thousand (1000) is *kilo;* the prefix for a thousandth (1/1000) is *milli.* One thousand meters (1000 m), therefore, equals one kilometer (1.0 km). One thousandth of meter (1/1000 m) equals one millimeter (1.0 mm). Comparing metric to English then:

> One millimeter (1.0 mm) = 1/1000 meter –.0397 inch
> One thousand millimeters (1000 mm) = 1.0 meter (1.0 m) = 3.281 feet
> One thousand meters (1000 m) = 1.0 kilometer (1.0 km) = 3281.0 feet

POCKET STEEL RULER

The pocket steel ruler is the easiest of all measuring tools to use. The *inch scale,* is six inches long and is graduated in 10ths and 100ths of an inch on one side and 32nds and 64ths on the other side.

The *metric scale* is 150 millimeters long (approximately six inches) and is graduated in millimeters and half millimeters on one side. Sometimes metric pocket steel rulers are graduated in 64ths of an inch on the other side.

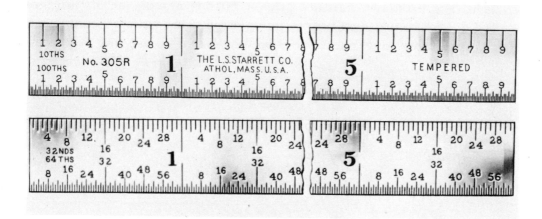

Inch pocket ruler

Metric pocket ruler

CALIPERS AND DIVIDERS

The illustration below show two types of calipers used in a machine shop. They are spring-loaded with a nut to lock measurements into position. The *inside caliper,* takes internal measurements, such as hole diameters and groove widths. The *outside caliper,* takes external measurements, such as rod diameters and stock thickness measurements. Shows a pair of dividers, a tool used to measure and transfer distances.

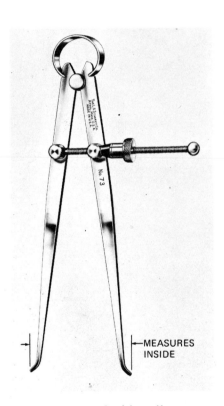

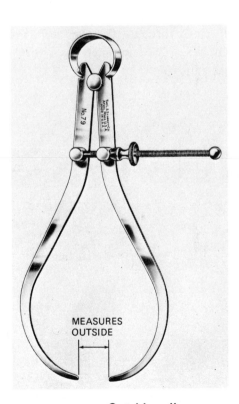

Inside caliper   Outside caliper

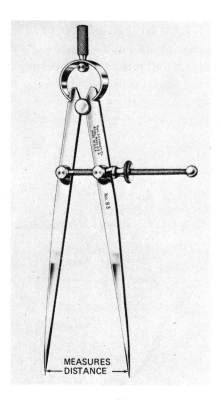

MEASURES
DISTANCE

Dividers

The calipers shown are not spring-loaded. They are more sensitive and care must be taken not to alter the measurement once it is made with the caliper.

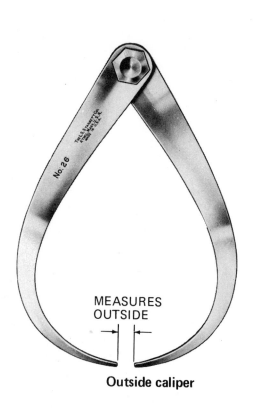

MEASURES
OUTSIDE

Outside caliper

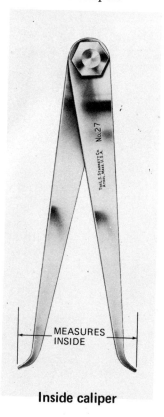

MEASURES
INSIDE

Inside caliper

MICROMETER

The *micrometer* is a highly accurate screw which rotates inside a fixed nut. As this screw is turned, it opens or closes a distance between the *measuring faces* of the anvil and spindle.

Place whatever is to be measured by the micrometer between the *anvil* and the *spindle*. Rotate the spindle by means of the *thimble* until the anvil and spindle come in contact with what is to be measured. The size is then determined by reading off the figures located on the *sleeve* and *thimble*.

Micrometers come in both English and metric graduations. They are manufactured with an English size range of 1 inch through 60 inches and a metric size range of 25 millimeters to 1500 millimeters. The micrometer is a very sensitive device and must be treated with extreme care.

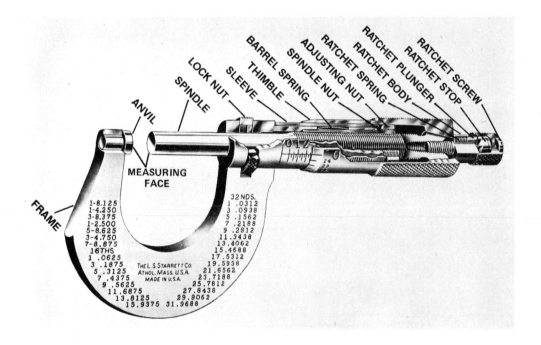

Micrometers come in various shapes and sizes.

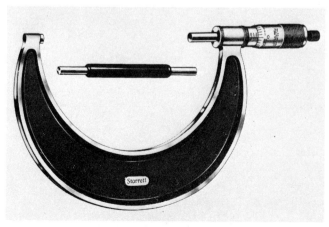

1″ TO 6″ MICROMETER

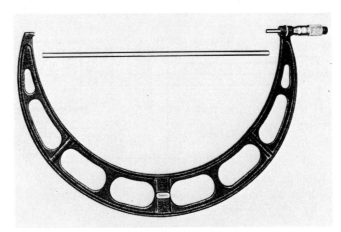

7″ TO 60″ MICROMETER

Measurements of inside surfaces are made with a *telescoping gauge.* The inside size is "locked," then the distance between ends is measured with a standard micrometer.

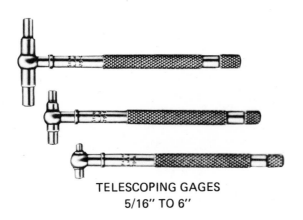

TELESCOPING GAGES
5/16″ TO 6″

The depth of holes, slots, and various projections are measured with a *micrometer depth gauge.*

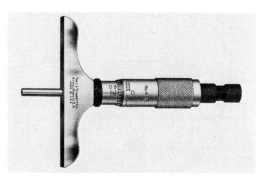

How to Read a Micrometer—English System

The sleeve is divided into 40 even spaces. This corresponds to the pitch of the spindle which is 40 threads per inch. One complete turn of the thimble moves the spindle 1/40 or .025 inch.

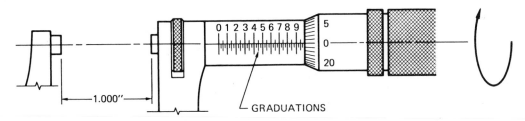

The beveled edge of the thimble is divided into 25 equal parts. Each line equals .001 inch.

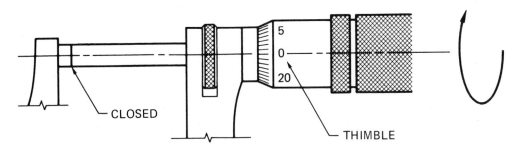

To read the micrometer: Multiply the number of vertical lines on the sleeve by .025 inch. Add the number of thousandths indicated on the sleeve.

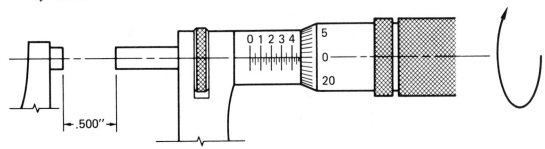

How to Read a Micrometer—Metric System

The sleeve is divided into 50 even spaces. This corresponds to the pitch of the spindle which is 0.5 millimeter. One full turn of the thimble moves the spindle 0.5 millimeter.

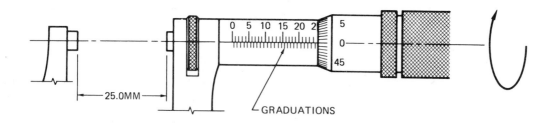

The beveled edge of the thimble is divided into 50 equal parts. Each line equals 1/50 of 0.5 millimeter or 0.01 millimeter.

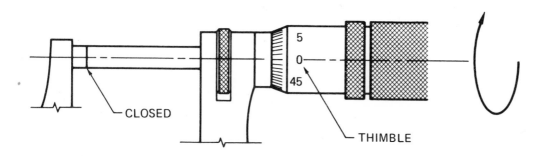

To read the micrometer: Multiply the number of vertical lines on the sleeve by 0.5 millimeter. Add the number of millimeters indicated on the sleeve.

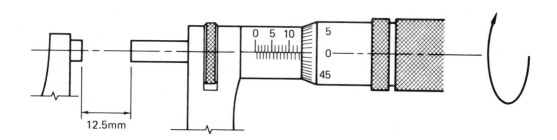

SAMPLE READINGS

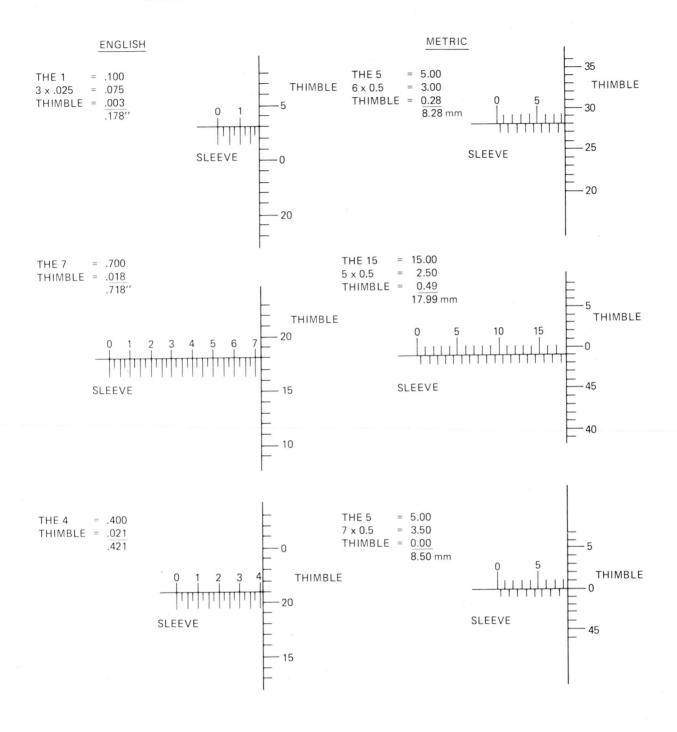

ENGLISH

THE 1   = .100
3 x .025   = .075
THIMBLE = .003
              .178"

METRIC

THE 5   = 5.00
6 x 0.5   = 3.00
THIMBLE = 0.28
              8.28 mm

THE 7   = .700
THIMBLE = .018
              .718"

THE 15   = 15.00
5 x 0.5   = 2.50
THIMBLE = 0.49
              17.99 mm

THE 4   = .400
THIMBLE = .021
              .421

THE 5   = 5.00
7 x 0.5   = 3.50
THIMBLE = 0.00
              8.50 mm

MICROMETER WORKSHEET

Calculate each reading below. See end of chapter for answers.

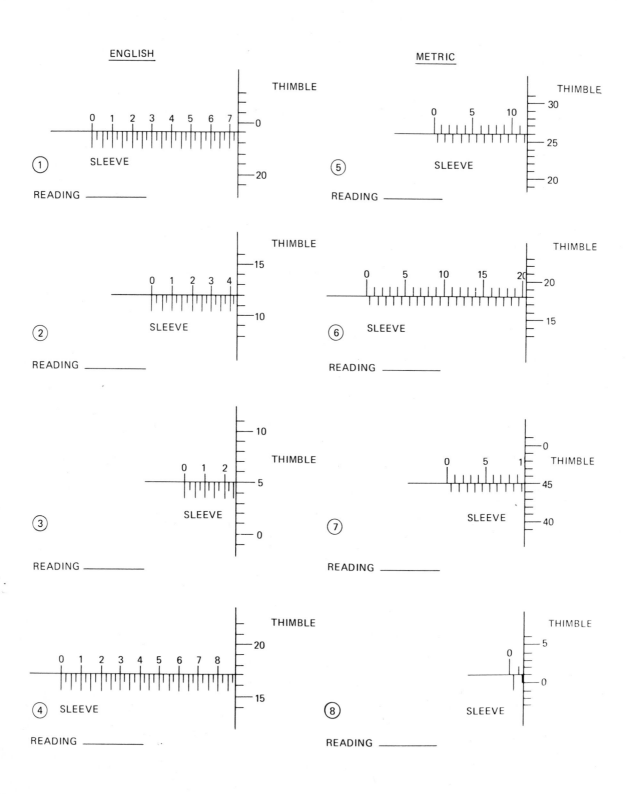

VERNIER CALIPERS

Vernier Calipers—English System

The vernier scale on the vernier caliper is read very much like a micrometer. The vernier reads to 1/1000 of an inch. The full inch is located on the main bar. Each inch is divided into 10 equal parts (1/10 inch) and in turn each tenth is divided into four equal parts. The vernier scale is divided into 25 equal parts, each representing .001 inch.

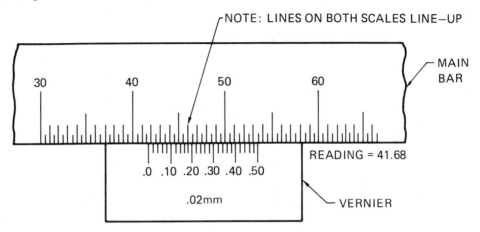

Using the above illustration, a reading is calculated as follows: From the main bar:

1. Record the full inches—1.000 (in thousandths)
2. Record the 1/10th inch—.400 (in thousandths)
3. Record the 1/25th inch—.025 (in thousandths)

From the vernier scale:

4. Count the number of graduations on the vernier scale from 0 to a line that coincides with a line on the main bar

$$\frac{-\ .011}{1.436}\ \text{(in thousandths)}$$
   inches

Vernier Calipers—Metric System

The vernier scale in the next illustration is a metric vernier and reads very much like the English vernier. It can be read to 0.02 millimeter. Each main bar graduation is in 0.5 millimeter with each 20th graduation numbered. The vernier is divided into 25 even parts, each representing 0.02 millimeter. To read the caliper, first record the total millimeters between zero on the main bar and zero on the vernier. Count the number of graduations on the vernier from zero to a line that coincides with a line on the main bar. Multiply that number times 0.02 millimeter and add this number to get the total reading.

Using the metric vernier scale, the vernier zero is 41.5 millimeters past the location of the zero on the main bar. Therefore:

$$41.5 \text{ mm} + (9 \times 0.02 \text{ mm}) = 41.68 \text{ mm total reading.}$$

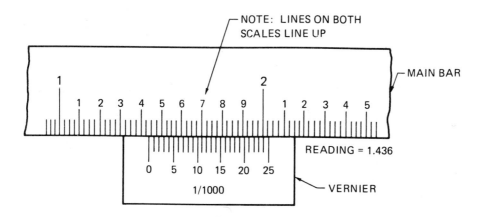

Vernier calipers have the ability to measure both outside an object and inside an object. When measuring an outside size, use the bottom scale. When measuring an inside size, use the top scale.

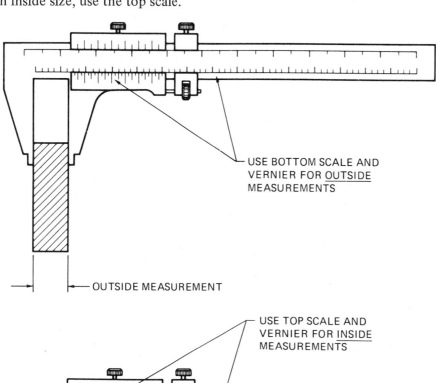

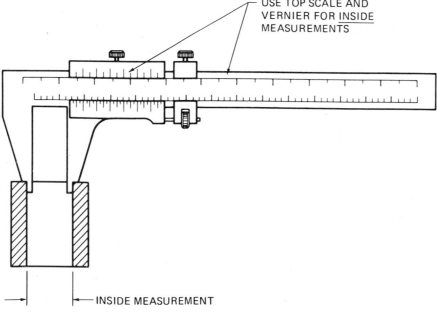

MICROMETER WORKSHEET ANSWERS

After practicing with the micrometer, most calculations can be made directly from the instrument without doing the math as illustrated.

(1) Reading .749"

The "7"   = .700
1 × .025   = .025
Thimble   = .024
              .749"

(2) Reading .437"

The "4"   = .400
1 × .025   = .025
Thimble   = .012
              .437"

(3) Reading .255"

The "2"   = .200
2 × .025   = .050
Thimble   = .005
              .255"

(4) Reading .892"

The "8"   = .800
3 × .025   = .075
Thimble   = .017
              .892"

(5) Reading 11.76 mm

The "10" = 10.00
3 × .5   = 1.50
Thimble   = 0.26
            11.76 mm

(6) Reading 20.18 mm

The "20" = 20.00
0 × 0   = 0.00
Thimble   = 0.18
            20.18 mm

(7) Reading 9.95 mm

The "5" = 5.00
9 × .5   = 4.50
Thimble   = 0.45
            9.95 mm

(8) Reading 1.51 mm

The "0" = 0.00
3 × .5   = 1.50
Thimble   = .01
            1.51 mm

EVALUATION

1. What are the standard lengths for English and metric pocket steel rulers?

   _____

   _____

   _____

   _____

2. How long is a metre as compared to the English system of measurement?

   _____

   _____

   _____

   _____

3. In the metric system each prefix represents equal steps of?

   _____

   _____

   _____

   _____

4. What countries are still using the 'English' system of measuring?

   _____

   _____

   _____

   _____

5. On a micrometer, which two parts are used to determine a dimension?

   _____

   _____

   _____

   _____

6. Which size micrometer in the metric system is equal to the one-inch size in the English system?

   _____

   _____

   _____

   _____

7. How is the sleeve graduated in the metric and English systems?

   _____

   _____

   _____

   _____

8. How is the thimble graduated in the metric and English systems?

   _____

   _____

   _____

   _____

9. What is the degree of accuracy possible with a vernier English caliper? A vernier metric caliper?

   _____

   _____

   _____

   _____

10. How many inches are there in 254 millimeters? Round off the answers to the nearest inch.

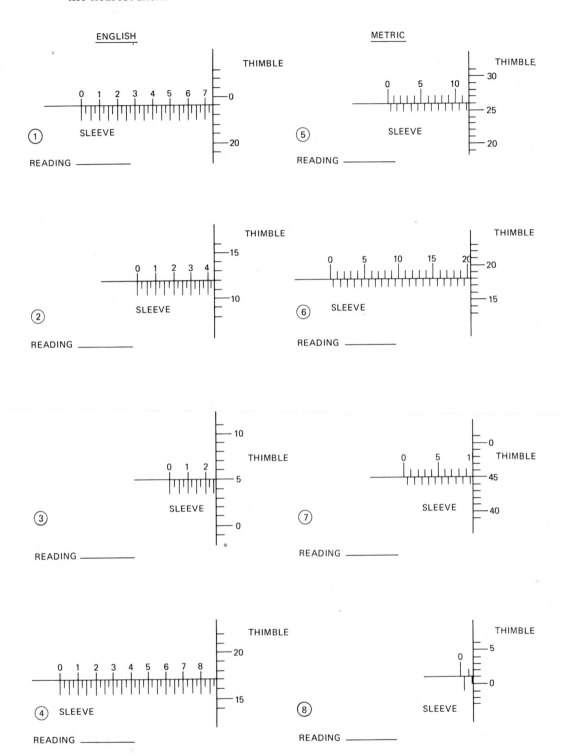

EVALUATION

Using the thimble/sleeve settings below "read" the sizes of the micrometer 1 through 8 in the spaces provided. See end of chapter for answers.

ANSWERS TO TERMS TO KNOW AND UNDERSTAND

Pocket Steel Ruler—Simplest measuring tool approx. 6 inches (150 mm) long/ graduated in 1/100 of an inch (or 1/2 millimeter).

English System—Using inches, feet, yards, etc.

Metric System—Using millimeters, meters, kilometers, etc.

Meter—A unit in the metric system and is equal to 3.281 feet.

Kilometer—1000 meters or 3281 feet.

Millimeter—1000 millimeters = 1 meter.

Calipers—A measuring device, can measure inside and outside surfaces.

Micrometer—Highly accurate measuring device.

> *Spindle*—The rotating measuring face.
>
> *Sleeve*—Where the measuring figures are located.
>
> *Thimble*—Where the measuring figures are located.
>
> *Frame*—Supports the complete assembly.
>
> *Anvil*—The fixed measuring face.

Telescoping Gauges—Used to measure inside surfaces—similar to a micrometer.

Depth Gauges—Used to measure depths—similar to a micrometer.

Vernier Calipers—Calipers with thumb-screw adjustments and a vernier.

Vernier—A method to read accurately between graduations.

Inside/outside Measurements (calipers)—An instrument that can measure both an inside dimension *and* an outside dimension.

MICROMETER READING

Your answers should agree with the answers below. After using the micrometer a while, calculations can be made directly from the micrometer without doing the math as outlined below.

English system:

Metric system:

①

The '7'	=	.700
1 × 0.25	=	.025
Thimble	=	.024
		.749″

⑤

The '10'	=	10.00
3 × .5	=	1.50
Thimble	=	0.26
		11.76 mm

②

The '4'	=	.400
1 × 0.25	=	.025
Thimble	=	.012
		.437″

⑥

The '20'	=	20.00
0 × 0	=	0.00
Thimble	=	0.18
		20.18 mm

③

The '2'	=	.200
2 × .025	=	.050
Thimble	=	.005
		.255″

⑦

The '5'	=	5.00
9 × .5	=	4.50
Thimble	=	0.45
		9.95 mm

④

The '8'	=	.800
3 × .025	=	.075
Thimble	=	.017
		.892″

⑧

The '0'	=	0.00
3 × .5	=	1.50
Thimble	=	.01
		1.51 mm

EVALUATION ANSWERS

English system: **Metric system:**

①

The '6'	=	.600
1 × .025	=	.025
Thimble	=	.024
		.649″

②

The 5	=	5.00
3 × .5	=	1.50
Thimble	=	0.16
		6.66 mm

③

The '3'	=	.300
1 × .025	=	.025
Thimble	=	.007
		.332″

④

The 15	=	15.00
0 × 0	=	0.00
Thimble	=	.013
		15.13 mm

⑤

The '1'	=	.100
2 × .025	=	.050
Thimble	=	.005
		.155″

⑥

The 5	=	5.00
9 × .5	=	4.50
Thimble	=	0.45
		9.95 mm

⑦

The '7'	=	.700
3 × .025	=	.075
Thimble	=	.012
		.787″

⑧

The 0	=	0.00
3 × .5	=	1.50
Thimble	=	.01
		1.51 mm

EVALUATION ANSWERS

1. A standard size pocket steel ruler is *6 inches (150 mm) long.*
2. A meter is *3.281 feet long.*
3. Prefixes represent steps of 10 equal spaces.
4. United States, Great Britain, and Canada use the "English" system of measuring.
5. The graduations on the *sleeve* and the graduations on the *thimble* are used to determine the dimension.
6. 25 mm micrometer is the equivalent to the 1.000 inch micrometer.
7. English system = .025 graduations/metric system = 0.5 mm graduations.
8. English system = .001 graduations/metric system = 0.01 mm graduations.
9. English system measurements to 1/1000 of an inch/metric system, measurements to .02 mm.
10. *10.0 inches* (.03937 x 254).

UNIT 10

CHANGE PROCEDURES AND MASTER PARTS LIST

Objective: To understand the change procedure used in industry and how to read and understand a master parts list.

TERMS TO KNOW AND UNDERSTAND

See end of chapter for answers.

Change Procedure _____

_____

_____

E.C.R. _____

_____

E.C.O. _____

_____

M.P.L. _____

_____

Assembly Drawing _____

_____

Sub-assembly Drawing _____

_____

Detail Drawing _____

_____

Purchased Parts _____

_____

_____

HOW A CHANGE IS MADE

Once a drawing has been issued, no one, including the draftsperson who drew the drawing can make a change on the original unless a strict change procedure is followed. Each company has its own method for change procedures. Outline below is one such method:

Step 1. Anyone throughout the company can suggest a change. Each suggested change is written out on an *E.C.R.* (Engineering Order *Request*) Form.

Step 2. Once a week a change committee, made up of various department heads, discuss each E.C.R. If the suggestion, improves the product, is a better manufacturing method or saves money the E.C.R. is approved.

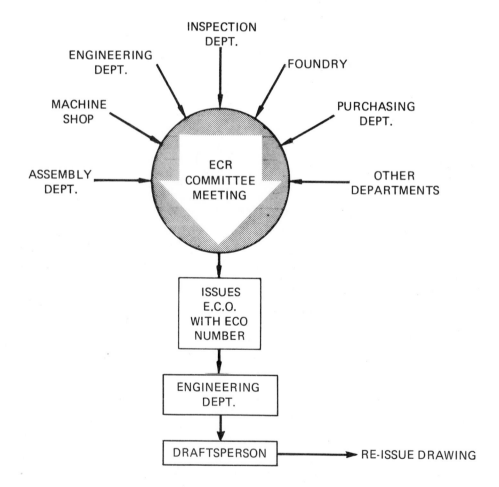

Step 3. An E.C.O. (Engineering Change Order) is issued to the Drafting Department.

Step 4. The drawing is changed or in the event of a major change, redrawn.

Step 5. The changed or new drawing is re-issued along with the E.C.O. Form describing exactly *what* was changed, and the reason the change was made.

CHANGE PROCEDURE

Various companies use different "standards" or rules governing changes, usually, the revision record box is located in the Title Block and includes the following:

1. Change letter (noted also beside the drawing number).

2. Date of change.

3. E.C.O. number.

4. Briefly what was changed (for example, "was .750"/"added 1/8 radii"/ "re-drawn"/etc.)

5. Name of detailer (draftsman) making the change.

It is *important* that all changes go through a definite procedure so all departments are notified of a change.

Note:

1. If the change is extensive, the drawing is completely redrawn and given the same drawing number, the obsolete original is *not* destroyed, it is marked "obsolete" and filed in the obsolete file.

2. If the change is fairly simple the original is carefully erased (so as not to damage the original) where the change is to be made and the change is carefully added to scale on the original.

3. In the event the change is *very minor,* such as a small dimension change, the dimension alone is changed and a wavy line is put under that dimension indicating "out-of-scale."

This is an example of a detail drawing with four changes (A, B, C, and D), note how each change was noted by its letter inside a baloon, and listed below in the revision record block. The E.C.R. number should also be included.

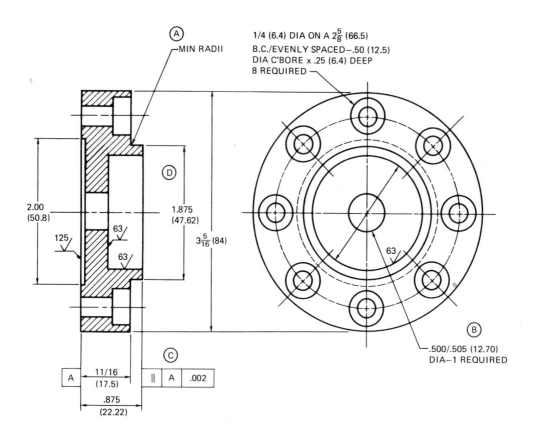

D	WAS 1.870 DIA		776987	PRS	RBC	9 APR 78
C	ADDED TOLERANCE		776953	JAN	RBC	10 JAN 78
B	WAS .375/.380 DIA		776914	CRA	PRS	8 DEC 77
A	ADDED NOTE		776872	JAN	RBC	15 NOV 77
REV.	CHANGE		E.C.O. NO.	BY	CHECKED	DATE
	REVISION RECORD					

KINDS OF DRAWINGS

There are various kinds of drawings used in a mechanical engineering department.

Design Layout Drawing

All major designs start from a design layout. This is usually a sketch drawing, full size or to scale. It is up to the drafter to draw each part depicted on this layout. In drawing the parts, the drafter must use the basic ideas in the design layout, but may change it to fit standard material stock, standard methods of manufacturing, and standard material sizes. If changes are made, they should be reviewed with the designer for his approval. Most of the time the designer draws parts as close to size as possible, but a design layout is not usually dimensioned unless particular dimensions must be maintained.

Assembly Drawing

Any product that has more than one part must have an assembly drawing. The assembly drawing shows how a product is assembled when completed. It can have one, two, three, or more views that are placed in the usual positions. One view is often a section view to illustrate the various parts and how they are assembled. Each part in an assembly drawing is identified by a circled detail number.

Permanently Fastened Parts Drawing

When two or more parts are permanently fastened such that they cannot be disassembled after assembly, they are called-out as in a sub-assembly drawing.

Detail Drawing

Each part must have its own fully dimensioned detail drawing, its own drawing number, and its own drawing title. All the information needed to manufacture the part is included in the detail drawing. The shape must be shown in the views given, features must be dimensioned and located, and specifications given on the drawing or title block.

Purchased Parts

A manufacturing company cannot afford to make standard items such as nuts, bots, and washers that can be purchased ready for use. A drafter should try to design around standard parts whenever possible. Such parts are not drawn but simply called-out on an assembly drawing by size, material, and finish. Other nonhardware parts may be treated in a similar fashion.

MASTER PARTS LIST

A master parts list is a list of all the parts required to put an assembly together. This list includes all detail drawings, all subassembly drawings; the assembly drawing and all purchased parts. The M.P.L. includes all drawing numbers and the required number of each part to complete one assembly. It is the responsibility of the draftsperson to make up the M.P.L. In order to fully understand a M.P.L., the craftsperson must know the various kinds of drawings used in industry. Carefully study each kind of drawing as described below:

INDEXING OF THE M.P.L.

A M.P.L. is laid out very much like an outline of a book would be done, i.e., important items first with related information indented below. In order of importance, 1. The *assembly drawing* (*one per* each M.P.L.) is the most important item thus is placed to the far left side of the description block. 2. Any *sub-assemblies* are indented one space to the right of the assembly drawing call-off. 3. *detail drawings* are indented two spaces to the right of the assembly drawing call-off. 4. *Purchase parts* are indented three spaces to the right of the assembly drawing call-off.

Note: Purchased parts do not have plan numbers thus, in the space provided for "part number", list it as "Purch." (purchased).

A M.P.L. usually follows this format:

1. Starts with the assembly drawing—first line, far *left*.

2. List all sub-assemblies, detail drawings and purchased parts *in the order would normally be assembled*.

3. Under each sub-assembly will be all parts indented, that are required to make up that sub-assembly.

4. Any miscellaneous parts are added at the end of the list—again, in the order they will be assembled.

For example:

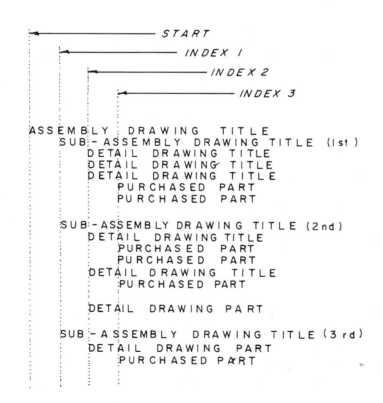

```
                              START
                        INDEX 1
                      INDEX 2
                    INDEX 3

ASSEMBLY  DRAWING   TITLE
   SUB-ASSEMBLY DRAWING TITLE  (1st)
      DETAIL  DRAWING TITLE
      DETAIL  DRAWING  TITLE
      DETAIL  DRAWING TITLE
         PURCHASED  PART
         PURCHASED  PART

   SUB-ASSEMBLY DRAWING TITLE (2nd)
      DETAIL  DRAWING TITLE
         PURCHASED  PART
         PURCHASED  PART
      DETAIL  DRAWING  TITLE
         PURCHASED PART

      DETAIL  DRAWING PART

   SUB-ASSEMBLY  DRAWING TITLE (3rd)
      DETAIL  DRAWING PART
         PURCHASED PART
```

Carefully study this M.P.L. Note what parts are sub-assemblies, which are detailed drawings, and which are purchased parts.

| | | PARTS LIST | | |
No.	Plan No.	Description	Material	Quan.
1	D-77942	VICE ASSEMBLY – MACHINE	AS NOTED	1
2				
3	C-77947	BASE – VICE	AS NOTED	1
4	B-77952	BASE – LOWER	C.I.	1
5	A-77951	BASE – UPPER	C.I.	1
6	A-77946	SPACER – BASE	STEEL	1
7	PURCH.	BOLT – ½ – 13 UNC – 2″ LG.	STEEL	1
8	PURCH.	NUT – ½ – 13 UNC	STEEL	1
9				
10	C-77955	JAW – SLIDING	STEEL	1
11				
12	A-77954	SCREW – VICE	STEEL	1
13	A-77953	ROD – HANDLE	STEEL	1
14	A-77956	BALL – HANDLE	STEEL	2
15				
16	A-77961	PLATE – JAN	STEEL	2
17	PURCH.	SCREW – ¼ – 20 UNC	STEEL	4
18				
19	A-77962	COLLAR	STEEL	1
20				
21	A-77841	KEY – SPECIAL VICE	STEEL	2
22				
23	PURCH.	BOLT – ½ – 13 UNC – 4″ LG.	STEEL	4
24	PURCH.	NUT – ½ – 13 UNC	STEEL	4

Company Name Company Address	Model No. 999	Parts Lister J.A.N.	Date 5 MAR. 79
Title VICE ASSEMBLY – MACHINE	Page 1 of 1 Pages		DWG No. D-77942

M.P.L. WORKSHEET

Using the sample M.P.L. (above,) answer the following questions. See end of chapter for answers.

1. How many parts are purchased? What line numbers are they? _____
    _____

2. How many sub-assemblies are there? What line numbers are they? _____
    _____

3. What is the general order of the parts (as *listed* on the MPL)? _____
    _____

4. How many detail parts are there? What line numbers are they? _____
    _____

5. What is the model number? What is the title? _____
    _____

6. What part numbers make up the screw-vice? How many parts are there to
    that sub-assembly? _____

7. What is the plan number for line number 17? _____
    _____

8. What "material" is used C-77947 sub-assembly? _____
    _____

9. What *size* paper is the *assembly drawing* drawn on? _____
    _____

EVALUATION

Answer all questions below. See end of chapter for answers.

1. What is an E.C.O.? (Explain in full) _____
    _____

2. Who makes up an E.C.R. committee? _____
    _____

3. What must be included on an E.C.O.? (List five things). _____
    _____

4. What is an assembly drawing? _____
    _____

5. Does an assembly drawing have any dimensions? _____
    _____

6. Why does a company purchase standard items? _____
    _____

7. What is a M.P.L.? (Explain in full). _____
    _____

8. What order are the parts listed in on a M.P.L.? _____
    _____

9. What number is usually used on a M.P.L.? _____

10. What is meant by "indexing of a M.P.L."? (Explain in full). _____
    _____
    _____

ANSWERS TO TERMS TO KNOW AND UNDERSTAND

Change Procedure—A standard set procedure used to make all changes through E.C.R.(s). Established by individual companies.

E.C.R.—Engineering Change *Request.*

E.C.O.—Engineering Change *Order.*

M.P.I.—Master Parts List (A list of *all* parts that make up an assembly. A complete unit).

Assembly Drawing—A master drawing showing all components in place (assembled) with all plan numbers noted.

Sub-assembly Drawing—An assembly of two or more parts that are assembled into a larger assembly.

Detail Drawing—A drawing showing all dimensions, finishes, notes, etc. of a single part.

Purchased Parts—Any part that is *not* manufactured by the company.

M.P.L. WORKSHEET ANSWERS

1. Five purchased parts (line numbers 7, 8, 17, 23, and 24). (14 total per assembly)
2. Two sub-assemblies (line numbers 3 and 12).
3. Place the parts in the order recommended or suggested for assembly.
4. Nine detail parts (line numbers 4, 5, 6, 10, 13, 14, 19, 21, and 16).
5. 999 model number, Vice Assembly—Machine.
6. (One) A-77953 and (two) A-77956—3 parts total.
7. No plan number—purchased parts are listed only, as "purch."
8. C.I. (base-lower) C.I. (base-upper) steel (spacer-base) and steel (bolt/nut) "as noted" indicates each *part* in the assembly is noted in its own call-off.
9. "D" size or 22 x 34 inches (560 mm x 865 mm).

EVALUATION ANSWERS

1. An E.C.O. is a Engineering change order, it gives permission to make a change on a particular drawing, *and* the reason why the change was made.
2. Usually a representative or head of each department in the company.
3. An E.C.O. must include:
    1. The change letter.
    2. Date of change.
    3. E.C.R. number.
    4. Briefly what was changed.
    5. Name of person making change.
4. An assembly drawing is a drawing of more than two parts combined with call-offs of what each part is and how many parts required of each.
5. Usually no, only for reference or general overall size. (All detail parts that make up the assembly have dimensions).
6. A company purchases parts because they can purchase them cheaper than they can make them.
7. A M.P.L. is a Master Parts List, it lists all the parts required to put together an assembly.
8. Parts are listed in the order they will be assembled together.
9. A M.P.L. usually uses the exact same number as the assembly drawing.
10. Indexing means: *Assembly Drawings* start on the far left side; *sub-assemblies* are indented 1 place in; *detail drawings* are indented 2 place in; *purchase parts* are indented 3 places in. This is done so one can see at a glance what parts make-up each sub-assembly; which parts are detail drawings and which parts are purchased.

ASSEMBLY DRAWING STUDY

Objective: To learn to interpret assembly drawings using, as a review, all drafting practices covered thus far.

M.P.L. SAMPLE

Using this M.P.L., the assembly drawing of an airplane engine, (page 322) and the various drawings provided, (pages 323–335), fill out this parts list.

PARTS LIST					
No.	Plan No.	Description		Material	Quan.
1	—	ENGINE – MODEL AIRPLANE		AS NOTED	1
2					
3					
4					
5					
6					
7					
8					
9					
10					
11					
12					
13					
14					
15					
16					
17					
18					
19					
20					
21					
22					
23					
24					

Company Name Company Address	Model No.	Parts Lister	Date
Title	Page ___ of ___ Pages		DWG. No.

A863515
HEAD-CYLINDER
1 REQ'D.

SCREW-FILL HD. MACH.
6-32 UNC-2A×5/16 LG.
3 REQ'D.

A863509
PIN-PISTON
1 REQ'D.

A863510
SPACER
2 REQ'D.

A863513
SLEEVE-CYLINDER
1 REQ'D.

A863511
BACKPLATE
1 REQ'D.

(SCREW-FILL. HD. MACH.
6-32 UNC - 2A × 7/16 LG.)
4 REQ'D.

A863514
PISTON
1 REQ'D.

SCREW FILL. HD. MACH
6-32 UNC-2A×1 3/16 LG
3 REQ'D.

A863512
FINS CYLINDER
1 REQ'D.

A863505
ROD-CONN.
1 REQ'D.

A863506
DRIVE WASHER
1 REQ'D.

A863508
PIN-DRIV.
1 REQ'D.

A863507
SPINNER
1 REQ'D.

A863504
CRANKCASE
1 REQ'D.

863503
CRANKCASE
SUB-ASSEMBLY
1 REQ'D.

DO NOT SCALE THIS DRAWING

TOLERANCE UNLESS OTHERWISE SPEC	
FRACTIONS	±.015
.XX	±.015
.XXX	±.005
.XXXX	±.0005

HEAT TREATMENT

SCALE FULL SIZE

DRAWN BY JAN	DATE 11/18/81
APPROVED BY JCM	DATE 11/19/81
CHECKED BY RCB	DATE 11/18/81
MATERIAL AS NOTED	

TITLE AIRPLANE ENGINE ASSEMBLY
MODEL: TITAN 60

GENEVRO MACHINE CO.
GARDEN GROVE
CALIFORNIA

A863516

LET	CHANGE	DATE

CHANGE NOTICE

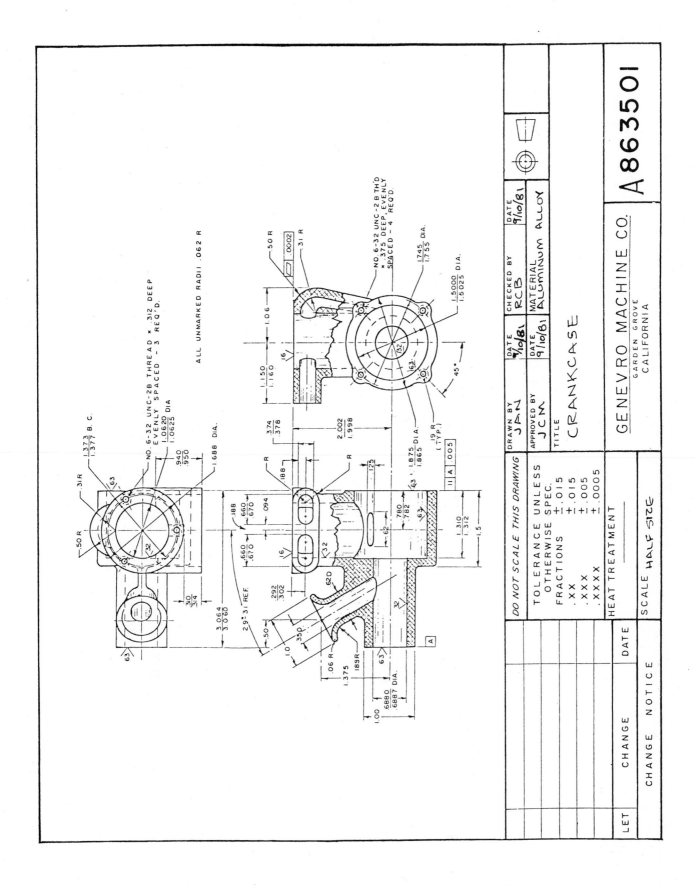

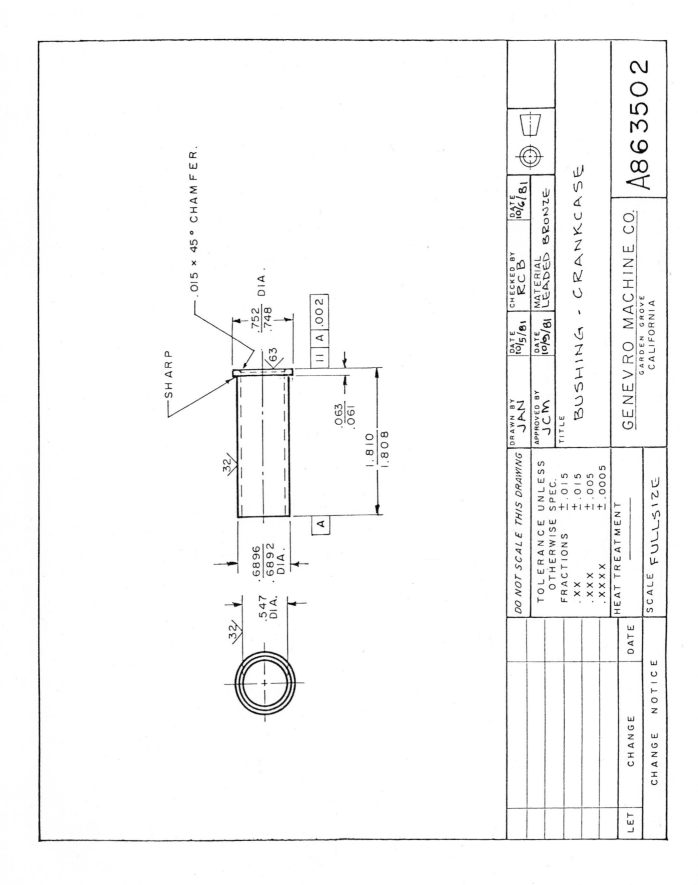

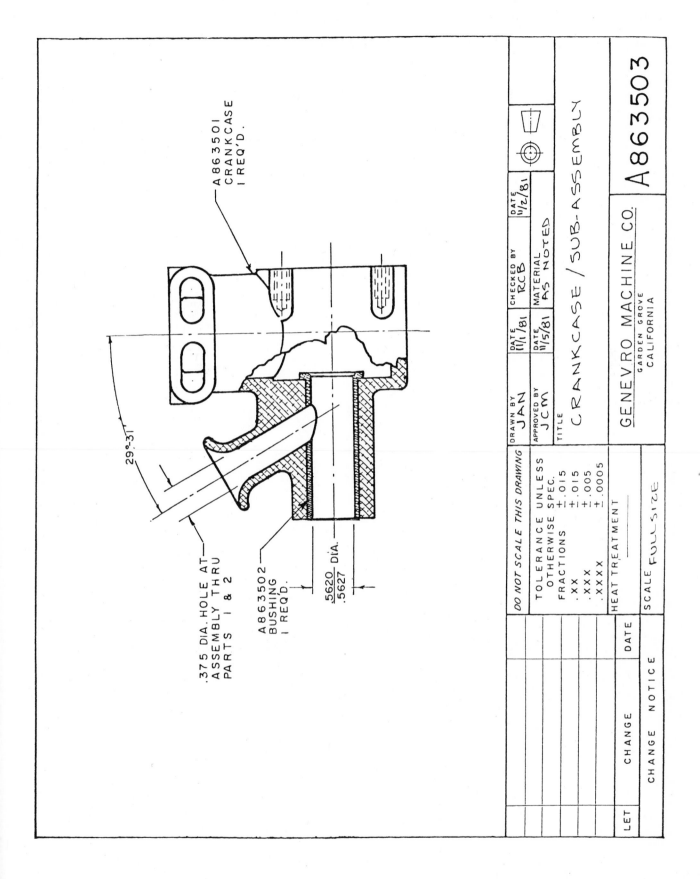

A 863501
CRANKCASE
I REQ'D.

.375 DIA. HOLE AT
ASSEMBLY THRU
PARTS I & 2

A863502
BUSHING
I REQ'D.

.5620 DIA.
.5627

29°-31'

DO NOT SCALE THIS DRAWING	DRAWN BY JAN	DATE 11/1/81	CHECKED BY RCB	DATE 11/2/81	
TOLERANCE UNLESS OTHERWISE SPEC.	APPROVED BY JCM	DATE 11/5/81	MATERIAL AS NOTED		
FRACTIONS ± .015	TITLE				
.XX ± .015					
.XXX ± .005					
.XXXX ± .0005		CRANKCASE / SUB-ASSEMBLY			
HEAT TREATMENT _____					
		GENEVRO MACHINE CO.			
SCALE FULL SIZE		GARDEN GROVE CALIFORNIA			A863503

LET	CHANGE	DATE

CHANGE NOTICE

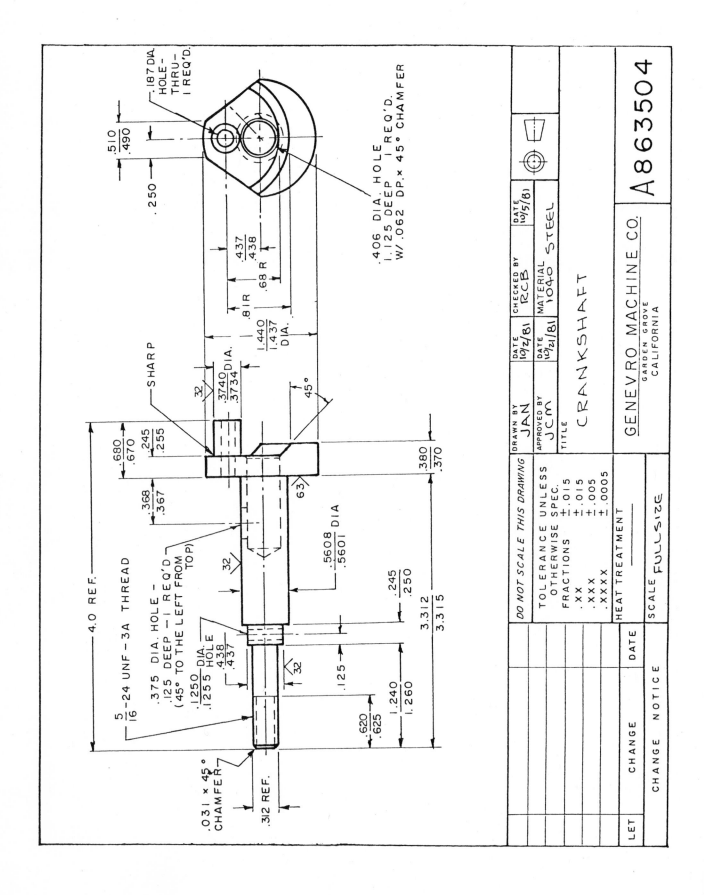

.187 DIA.
HOLE-
THRU-
I REQ'D.

.510
.490

.250

.406 DIA. HOLE
1.125 DEEP I REQ'D.
W/.062 DP.x 45° CHAMFER

.437
.438

.68 R

.81 R

1.440
1.437
DIA.

SHARP

.3740 DIA.
.3734

32

45°

.680
.670

.245
.255

.368
.367

32

.5608 DIA
.5601

63

.380
.370

4.0 REF.

5/16-24 UNF-3A THREAD

.375 DIA. HOLE-
.125 DEEP-I REQ'D
(45° TO THE LEFT FROM
TOP)

.1250 DIA.
.1255 HOLE .438
.437

32

.245
.250

3.312
3.315

.125

32

.620
.625

1.240
1.260

.031 x 45°
CHAMFER

.312 REF.

DO NOT SCALE THIS DRAWING

DRAWN BY
JAN

DATE
10/2/81

CHECKED BY
RCB

DATE
10/5/81

APPROVED BY
JCM

DATE
10/21/81

MATERIAL
1040 STEEL

TOLERANCE UNLESS
OTHERWISE SPEC.
FRACTIONS ±.015
.XX   ±.015
.XXX   ±.005
.XXXX   ±.0005

HEAT TREATMENT

SCALE FULLSIZE

TITLE
CRANKSHAFT

GENEVRO MACHINE CO.
GARDEN GROVE
CALIFORNIA

A863504

LET   CHANGE   DATE

CHANGE NOTICE

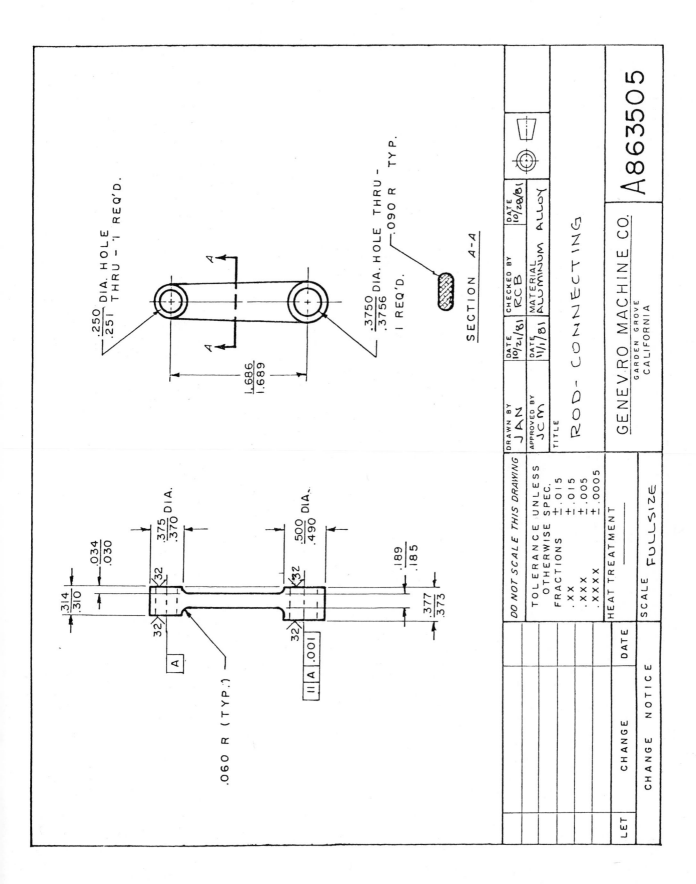

.250/.251 DIA. HOLE THRU – I REQ'D.

.3750/.3756 DIA. HOLE THRU – I REQ'D.

.090 R TYP.

SECTION A-A

1.686/1.689

.034/.030

.375/.370 DIA.

.500/.490 DIA.

.189/.185

.314/.310

.377/.373

32

32

A

.060 R (TYP.)

II A .001

DO NOT SCALE THIS DRAWING

TOLERANCE UNLESS
OTHERWISE SPEC.
FRACTIONS ±.015
.XX ±.015
.XXX ±.005
.XXXX ±.0005

HEAT TREATMENT _____

SCALE FULLSIZE

DRAWN BY JAN   DATE 10/21/81

CHECKED BY RCB   DATE 10/26/81

APPROVED BY JCM   DATE 11/1/81

MATERIAL ALUMINUM ALLOY

TITLE

ROD- CONNECTING

GENEVRO MACHINE CO.
GARDEN GROVE
CALIFORNIA

A863505

CHANGE NOTICE

LET   CHANGE   DATE

45° REF.

.040
.050

1.005
.995

.255
.245

.620
.630

.130
.120

1.190
1.180

.125
.126

.062

.437 DIA.
.438

.552
.562

DO NOT SCALE THIS DRAWING

TOLERANCE UNLESS
OTHERWISE SPEC.
FRACTIONS ±.015
.XX   ±.015
.XXX   ±.005
.XXXX   ±.0005

HEAT TREATMENT

SCALE FULLSIZE

DRAWN BY
JAN

DATE
9/15/81

APPROVED BY
JCM

DATE
9/21/81

CHECKED BY
RCB

DATE
9/18/81

MATERIAL C1018 STEEL

TITLE
DRIVE WASHER

GENEVRO MACHINE CO.
GARDEN GROVE
CALIFORNIA

A863506

LET | CHANGE | DATE
CHANGE NOTICE

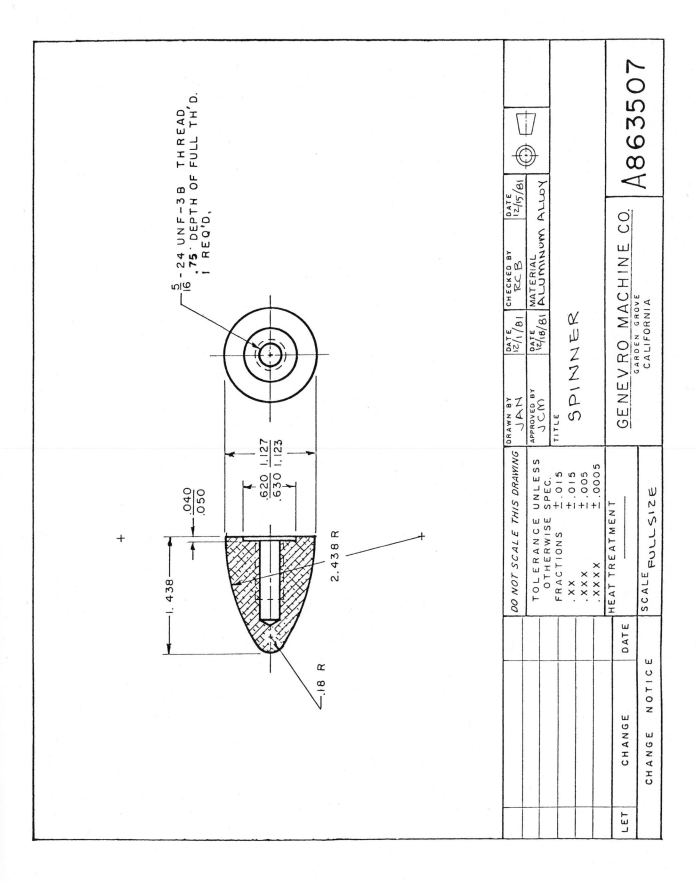

5/16 - 24 UNF - 3B THREAD
.75 DEPTH OF FULL TH'D.
1 REQ'D.

.040
.050

1.438

1.127
1.123

.620
.630

2.438 R

.18 R

DO NOT SCALE THIS DRAWING

TOLERANCE UNLESS
OTHERWISE SPEC.
FRACTIONS ±.015
.XX ±.015
.XXX ±.005
.XXXX ±.0005

HEAT TREATMENT _____

SCALE FULL SIZE

DRAWN BY JAN | DATE 12/1/81
APPROVED BY JCM | DATE 12/18/81

CHECKED BY RCB | DATE 12/15/81
MATERIAL ALUMINUM ALLOY

TITLE
SPINNER

GENEVRO MACHINE CO.
GARDEN GROVE
CALIFORNIA

A863507

CHANGE | DATE
CHANGE NOTICE
LET

$\dfrac{.252}{.253}$

$\dfrac{.314}{.310}$

$\dfrac{.150}{.140}$

$\dfrac{.310}{.320}$ R

DO NOT SCALE THIS DRAWING

DRAWN BY JAN

DATE 11/4/81

CHECKED BY RCB

DATE 11/14/81

APPROVED BY JCM

DATE 11/15/81

MATERIAL BRASS

TOLERANCE UNLESS OTHERWISE SPEC.
FRACTIONS ±.015
.XX   ±.015
.XXX   ±.005
.XXXX   ±.0005

HEAT TREATMENT

SCALE DOUBLE SIZE

TITLE

SPACER - PISTON

GENEVRO MACHINE CO.
GARDEN GROVE
CALIFORNIA

A8635 10

LET   CHANGE   DATE

CHANGE   NOTICE

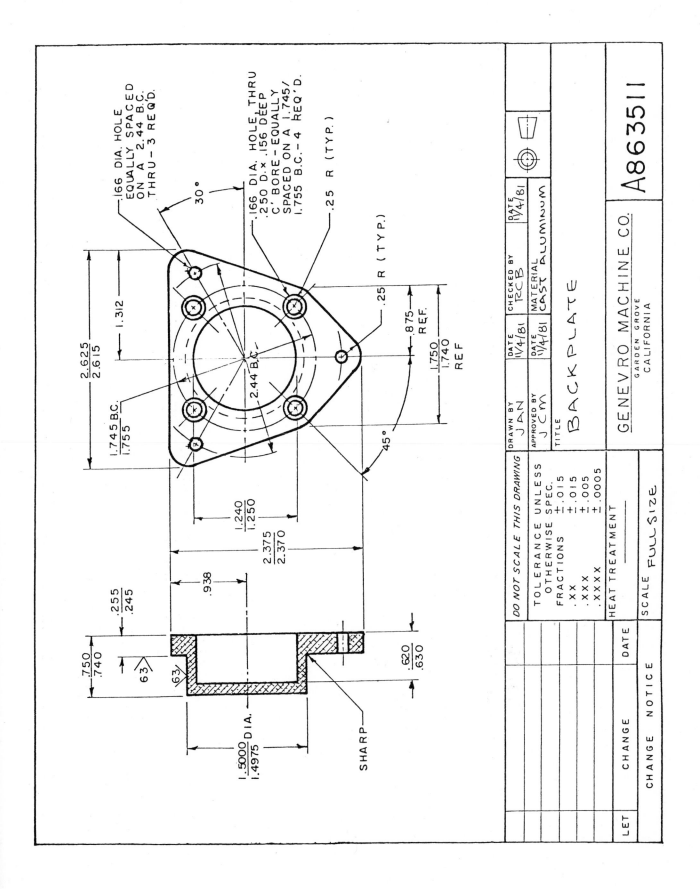

.166 DIA. HOLE
EQUALLY SPACED
ON A 2.44 B.C.
THRU–3 REQ'D.

30°

.166 DIA. HOLE THRU
.250 D.×.156 DEEP
C' BORE–EQUALLY
SPACED ON A 1.745/
1.755 B.C.–4 REQ'D.

.25 R (TYP.)

1.312

.25 R (TYP.)

2.625
2.615

.875
REF.

1.745 B.C.
1.755

2.44 B.C.

1.750
1.740
REF

45°

1.240
1.250

2.375
2.370

.938

.255
.245

.750
.740

63⁄

63⁄

.620
.630

1.5000 DIA.
1.4975

SHARP

DO NOT SCALE THIS DRAWING	DRAWN BY JAN	DATE 11/4/81	CHECKED BY RCB	DATE 11/4/81
TOLERANCE UNLESS OTHERWISE SPEC. FRACTIONS ±.015 .XX ±.015 .XXX ±.005 .XXXX ±.0005	APPROVED BY JCM	DATE 11/4/81	MATERIAL CAST ALUMINUM	

TITLE
BACKPLATE

GENEVRO MACHINE CO.
GARDEN GROVE
CALIFORNIA

A863511

HEAT TREATMENT ———

SCALE FULL SIZE

CHANGE NOTICE

LET	CHANGE	DATE

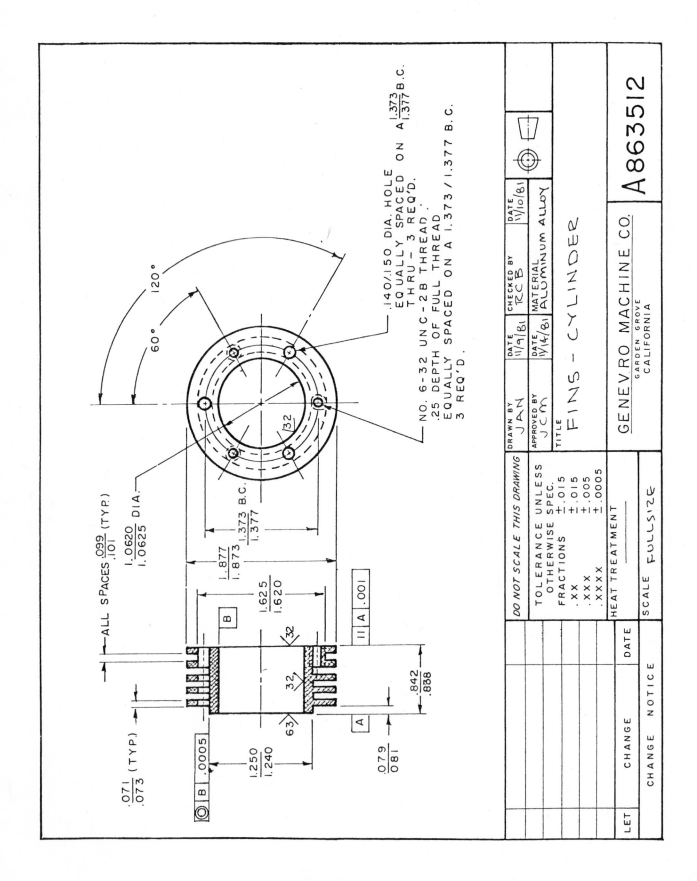

.140/.150 DIA. HOLE
EQUALLY SPACED ON A 1.373 B.C.
1.377
THRU – 3 REQ'D.

NO. 6-32 UNC-2B THREAD.
.25 DEPTH OF FULL THREAD.
EQUALLY SPACED ON A 1.373/1.377 B.C.
3 REQ'D.

120°

60°

32

ALL SPACES .099 (TYP.)
                .101

1.0620 DIA.
1.0625

1.373 B.C.
1.377

1.877
1.873

1.625
1.620

B

32

32

63

A

‖ A .001

.842
.838

.071 (TYP.)
.073

B .0005

1.250
1.240

.079
.081

DO NOT SCALE THIS DRAWING	DRAWN BY JAN	DATE 11/9/81	CHECKED BY RCB	DATE 11/10/81		
TOLERANCE UNLESS OTHERWISE SPEC.	APPROVED BY JCM	DATE 11/14/81	MATERIAL ALUMINUM ALLOY			
FRACTIONS ±.015	TITLE					
.XX ±.015	FINS – CYLINDER					
.XXX ±.005						
.XXXX ±.0005	GENEVRO MACHINE CO.			A863512		
HEAT TREATMENT	GARDEN GROVE CALIFORNIA					
SCALE FULLSIZE						

LET	CHANGE	DATE

CHANGE NOTICE

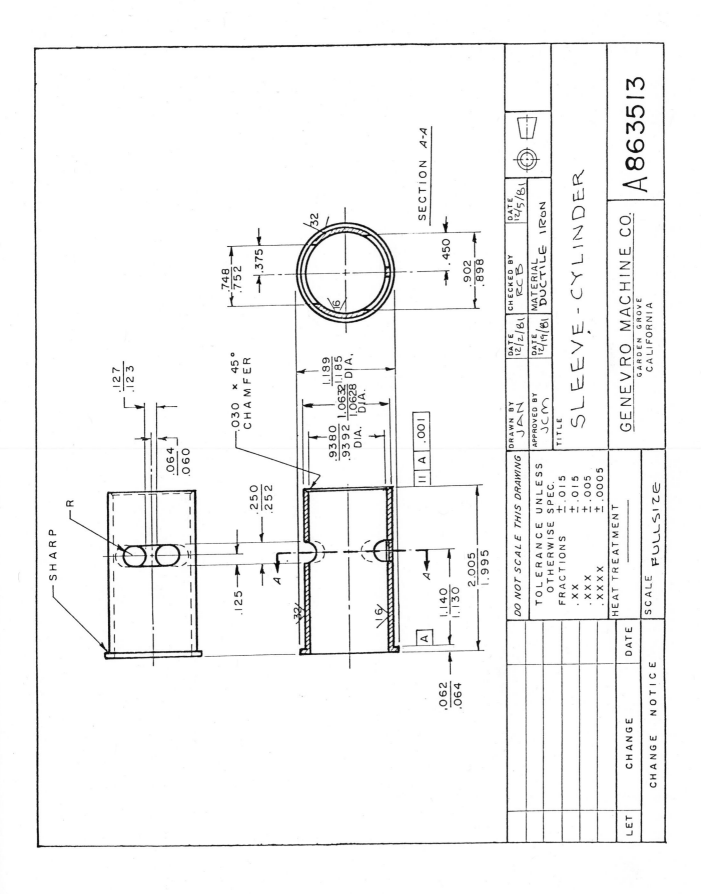

SECTION A-A

SHARP

R

.127
.123

.064
.060

.030 × 45°
CHAMFER

.250
.252

.125

.062
.064

.748
.752

.375

32

16

.902
.898

1.189
1.185 DIA.

1.0632
1.0628 DIA.

.9380
.9392 DIA.

32

16

A

A

2.005
1.995

1.140
1.130

| // | A | .001 |

| DO NOT SCALE THIS DRAWING | DRAWN BY JAN | DATE 12/2/81 | TITLE |
| TOLERANCE UNLESS OTHERWISE SPEC. | APPROVED BY JCM | DATE 12/9/81 | SLEEVE, CYLINDER |

CHECKED BY RCB   DATE 12/5/81

MATERIAL DUCTILE IRON

TOLERANCE UNLESS OTHERWISE SPEC.
FRACTIONS ±.015
.XX ±.015
.XXX ±.005
.XXXX ±.0005

HEAT TREATMENT

SCALE FULLSIZE

GENEVRO MACHINE CO.
GARDEN GROVE
CALIFORNIA

A 863513

LET | CHANGE | DATE

CHANGE NOTICE

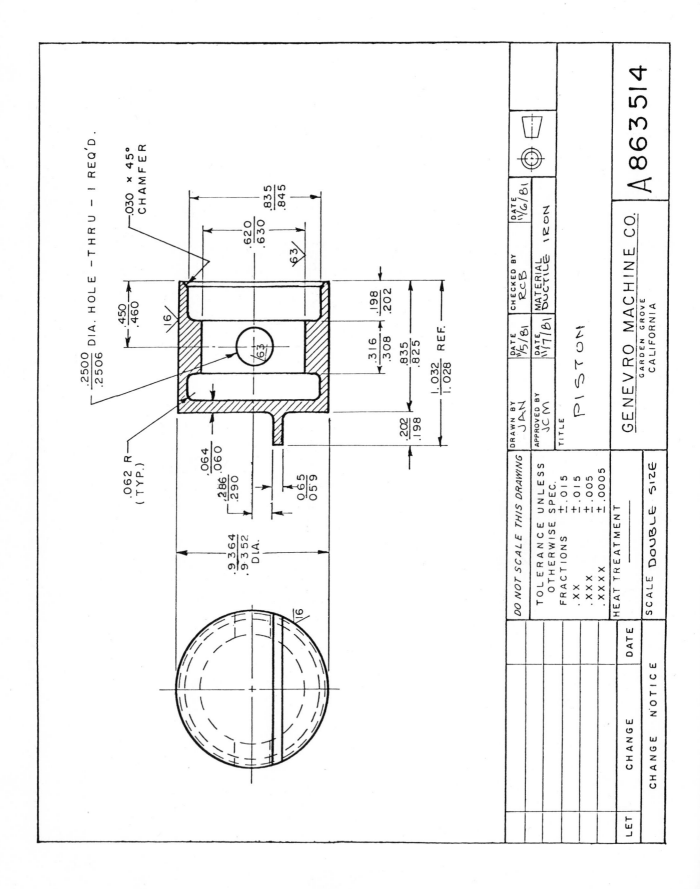

.2500 DIA. HOLE – THRU – 1 REQ'D.
.2506

.030 × 45° CHAMFER

.620
.630

.835
.845

63

.450
.460

16

63

.198
.202

.316
.308

.835
.825

1.032 REF.
1.028

.202
.198

.062 R
(TYP.)

.064
.060

.286
.290

.065
.059

.9364
.9352
DIA.

16

DO NOT SCALE THIS DRAWING

TOLERANCE UNLESS
OTHERWISE SPEC.
FRACTIONS ±.015
.XX ±.015
.XXX ±.005
.XXXX ±.0005

HEAT TREATMENT

SCALE DOUBLE SIZE

DRAWN BY
JAN
DATE
11/5/81

APPROVED BY
JCM
DATE
11/7/81

TITLE PISTON

CHECKED BY
RCB
DATE
11/6/81

MATERIAL
DUCTILE IRON

GENEVRO MACHINE CO.
GARDEN GROVE
CALIFORNIA

A863514

LET | CHANGE | DATE

CHANGE NOTICE

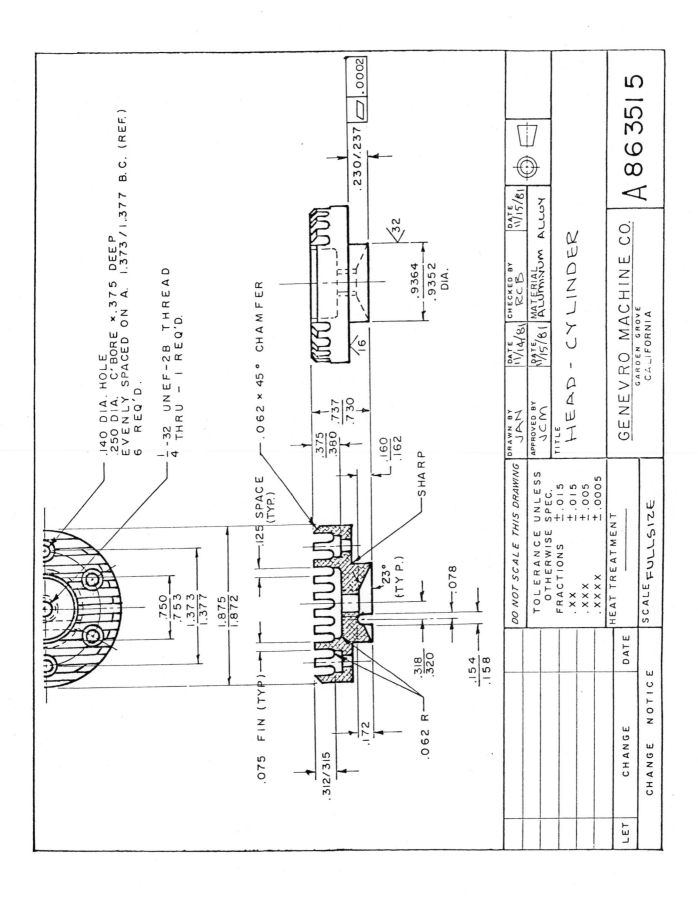

.140 DIA. HOLE
.250 DIA. C'BORE ×.375 DEEP
EVENLY SPACED ON A. 1.373/1.377 B.C. (REF.)
6 REQ'D.

¼-32 UNEF-2B THREAD
THRU - 1 REQ'D.

.062 × 45° CHAMFER

.125 SPACE (TYP.)

.075 FIN (TYP)

.750
.753

1.373
1.377

1.875
1.872

.312/.315

.318
.320

.154
.158

.172

.062 R

23° (TYP.)

SHARP

.078

.737
.730

.375
.380

.160
.162

.230/.237

| | .0002 |

32

.9364
.9352
DIA.

16

DO NOT SCALE THIS DRAWING

TOLERANCE UNLESS
OTHERWISE SPEC.
FRACTIONS ±.015
.XX   ±.015
.XXX ±.005
.XXXX ±.0005

HEAT TREATMENT

DRAWN BY
JAN

DATE
11/14/81

CHECKED BY
RCB

DATE
11/15/81

APPROVED BY
JCM

DATE
8/15/81

MATERIAL
ALUMINUM ALLOY

TITLE
HEAD - CYLINDER

GENEVRO MACHINE CO.
GARDEN GROVE
CALIFORNIA

A863515

SCALE FULLSIZE

LET   CHANGE   DATE

CHANGE NOTICE

REVIEW WORKSHEET

Using the completed M.P.L. provided below, answer the following questions. . See end of chapter for answers.

1. How many *sub-assemblies* are there in this assembly? _____

2. Is the backplate A863511 assembled with the main assembly *before* or *after* the cylinder sleeve A863513? Explain. _____ _____

3. What parts make up the crankcase sub-assembly? _____ _____

4. What material is used to mfg. part no. A863509? _____ _____

5. How many spacer rods are used? _____

6. Explain in full what the word "purch." means. _____ _____

7. What does item No. 13 mean? Explain in full. _____ _____ _____

8. How many screws hold the backplate (A863511) in place? How *long* are they? _____ _____

9. How many screws hold the cylinder head (A863515) in place? What length are they? _____ _____ _____

10. Where is the M.P.L. number usually derived from? _____ _____

M.P.L. SAMPLE ANSWERS

Your answer should be indented as noted below, be complete, including Title Block. (It does *not* have to be listed in the exact order.)

MASTER PARTS LIST				
NO.	PLAN NO.	DESCRIPTION	MATERIAL	QUAN.
1	A863516	ENGINE-MODEL AIRPLANE	AS NOTED	1
2	A863503	CRANKCASE SUB-ASSEMBLY	AS NOTED	1
3	A863501	CRANKCASE	ALUM·	1
4	A863502	BUSHING-CRANKCASE	LEAD/BRO.	1
5				
6	A863504	CRANKSHAFT	STEEL	1
7	A863505	ROD-CONNECTING	ALUM·	1
8	A863508	PIN-DRIVEWASHER	STEEL	1
9	A863509	PIN-PISTON	STEEL	1
10	A863510	SPACER-ROD	BRASS	2
11	A863511	BACKPLATE	ALUM.	1
12	PURCH.	SCREW-FILL. HD. MACH.	—	—
13	—	6-32 UNC-2A ×7/16 LG.	STEEL	4
14				
15	A863513	SLEEVE-CYLINDER	IRON	1
16	A863512	FINS-CYLINDER	ALUM.	1
17	A863514	PISTON	IRON	1
18	A863515	HEAD CYLINDER	ALUM.	1
19	PURCH.	SCREW-FILL. HD. MACH.	—	—
20	—	6-32 UNC-2A × 1 3/16 LG.	STEEL	3
21	PURCH.	SCREW-FILL.HD. MACH.	—	—
22	—	6-32 UNC-2A × 5/16 LG.	STEEL	3
23	A863506	DRIVEWASHER	STEEL	1
24	A863507	SPINNER	ALUM.	1

GENEVRO MACHINE CO. Garden Grove, Calif.	Model no. TITAN 60	Lister: NELSON	Date: 15 NOV 81
Title: MODEL AIRPLANE ENGINE		PAGE 1 OF 1 PAGES	DWG. NO A863516

EVALUATION

Answer the following questions. See end of chapter for answers.

1. Refer drawing A863501 (crankcase), what is the maximum overall height? _____

2. Explain in full the meaning of the geometric tolerancing symbols. (Refer drawing A863502, Busing-crankcase). _____

3. Refer drawing A863503 (crankcase sub-assembly), why is the .375 dia. hole drilled out now? _____

4. Refer drawing A863504 (crankshaft), at what angle is the .375 dia. hole drilled. _____

5. Refer drawing A863505 (rod-connecting), what kind of section view is section A-A? _____

6. Refer drawing A863506, (drivewasher), what is the 1/8 nom. size, slot called? _____

7. Refer drawing A863507 (spinner), how deep do the full threads go into the part? _____

8. Refer drawing A863508 (pin-drivewasher), explain what the 63 located in a "V" means. _____

9. Refer drawing A863509 (pin-piston), what machining process is done at each end? _____

10. Refer drawing A863510 (spacer-piston), what is the tolerance of the O.D.? _____

11. Refer drawing A863511 (backplate), what does "B.C." mean? _____

12. Refer drawing A863512 (fins-cylinder), what does the geometric symbol indicate? _____

13. Refer drawing A863513 (sleeve-cylinder), what is the surface finish on the O.D. and on the I.D.? _____

14. Refer drawing A863514 (piston), what material is used to manufacture this part? _____

15. Refer drawing A863515 (head-cylinder), what is the largest allowable size for the fins? _____

16. What is the *maximum* heighth of the engine assembly (from lowest point to the highest point)? _____

EVALUATION

Before beginning, review each term before starting. Answer each question. See end of chapter for answers.

1. Refer drawing A863501 (crankcase), what are the upper and lower limits of the I.D. cyliner wall? _____

2. Refer drawing A863502 (bushing-crankcase), what is the *tolerance* of the overall length? _____

3. Refer drawing A863503 (crankcase/sub-assembly), what is the M.M.C. of the hole diameter in the bushing? _____

4. Refer drawing A863504 (crankshaft), what is the M.M.C. of the largest diameter shaft? _____

5. Refer drawing A863505 (rod connecting), what is the *longest* distance the two holes can be apart and still be within tolerance? _____

6. Refer drawing A863506 (drivewasher), what is the *minimum* I.D. and O.D.? _____

7. Refer drawing A863507 (spinner), what is the upper and lower limits for the overall length? _____

8. Refer drawing A863508 (pin-drivewasher), what is the *allowance* of the drive-washer pin when it is in the final assembly? _____

9. Refer drawing A863509 (pin-piston), what is the *clearance* of the piston pin when it is in the final assembly? _____

10. Refer drawing A863510 (spacer-piston), what is the *allowance* of the piston spacer when it is in the final assembly? _____

11. Refer drawing A863511 (backplate), what is the *allowance/clearance* of the backplate when it is in the final assembly? _____

12. Refer drawing A863512 (fins-cylinder), what is the *allowance/clearance* of the cylinder fins to the cylinder sleeve? _____

REVIEW WORKSHEET ANSWERS

1. One sub-assembly (item No. 2).
2. It is suggested that the backplate A863511 be assembled *before* the cylinder sleeve A863513 because it is *listed* ahead of the latter.
3. The parts that make up the crankcase sub-assembly are A863501 crankcase and A863502 bushing-crankcase.
4. Piston pin A863509 material is *steel.*
5. There are two (2) spacer rods used.
6. "Purch." means that particular part or parts are to be *purchased* already made, not manufactured by the company assemblying the component.
7. Item No. 13: 6-32 UNC-2A—7/16 lg. means—(#6) dia. (.138)/(32) = threads per inch/(UNC) = Unified National *Coarse* Threads/(2) = Medium class of fit/(A) = External threads/7/16 inch length of screw.
8. Four screws hold the backplate in place—they are 7/16″ long.
9. Six screws hold the cylinder head in place. Three are 1 3/16 inch long and three are 5/16 inch long.
10. The M.P.L. number is usually derived from the assembly drawing number.

EVALUATION ANSWERS

1. Height = 2.002 + (1/2 x 1.875) = 2.9695
2. The right end must be parallel to the left end (A) within .002.
3. The hole is drilled *after* assembly in order to assure the .375 dia. hole will intersect the main shaft at exactly the correct position.
4. 45° to the *left* from top ctr. See note and hidden lines (right view).
5. *Removed* section.
6. A keyway.
7. The full threads go .75 deep into the part.
8. The 63 inside a "V" indicates that surface is to be finished or smoothed to 63 microinches.
9. 45° x .130 chamfer.
10. O.D. Tolerance = .314 – .310 = *.004.*
11. B.C. means bolt circle or the diameter of which various holes are located on.
12. The geometric symbol indicates this surface must be concentric to surface 'B' within .0005.
13. O.D. surface = 32 microinches. – I.D. surface = 16 microinches.
14. Material = Ductile iron.
15. Maximum fin size = .075 + .005 = *.080.*
16. 2.375 max. limit – .933 min. limit = *1.442* max. allowed limit (A863511)
    Plus
    2.002 max limit (A863501)
    Plus
    .842 max. limit (A863512)
    Plus
    .063 max. limit (A863502)
    Plus
    .737 max. limit – .230 min. limit = *.507* max. allowed limit (A863515)
    Equals – *4.856 max. overall heighth.*

EVALUATION ANSWERS

1. I.D. cylinder wall = 1.0620 *lower limit* and 1.0625 *upper limit*.

2. Overall length *tolerance* = .002 *tolerance*.

3. M.M.C. of the hole = *.5620* dia. (smallest hole size).

4. M.M.C. of the largest shaft = *.5608* dia. (largest limit for shaft dia.)

5. Longest distance = *1.689* (upper limit).

6. Minimum inside diameter = *.437*/minimum outside diameter = *1.190*.

7. Overall length = *1.443* upper limit/*1.433* lower limit.

8. Allowance = *.0*

    .1250 M.M.C. of hole (lower limit) A863504
    -.1250 M.M.C. of shaft (upper limit) A863508
    =.0000 Allowance

9. Clearance = *.0004 interference (press) fit*

    -.2510 *Smallest* shaft -A863509
    .2506 Largest hole -A863514
    =.0004 Clearance

    Note: Because the *shaft* is larger than the *hole*, there will be interference, thus a *press* fit.

10. Allowance = *.0006*

    .2520 M.M.C. Hole (lower limit) A863510
    -.2514 M.M.C. Shaft (upper limit) A863509
    =.0006 Allowance

11. Allowance = *.0*
    Clearance = *.0050*

    1.5000 M.M.C. Hole (lower limit) A863501
    -1.5000 M.M.C. Shaft (upper limit) A863511
    = .0000 Allowance

    1.5025 Largest hole (A863501)
    -1.4975 Smallest shaft (A863511)
    = .0050 Clearance

12. Allowance = *.0012* interference (Press) fit.
    Clearance = *.0003*

    Note: Because the *shaft* is larger than the *hole*, there will be interference thus a *press* fit.

    1.0632 M.M.C. Shaft (upper limit) A863513
    -1.0620 M.M.C. Hole (lower limit) A863512
    =0.0012 Allowance

    1.0628 Smallest shaft (lower limit) A863513
    -1.0625 Largest hole (upper limit) A863512
    = .0003 Clearance

APPENDIX A

INCH/METRIC — EQUIVALENTS					
Fraction	**Decimal Equivalent**		**Fraction**	**Decimal Equivalent**	
	Customary (in.)	Metric (mm)		Customary (in.)	Metric (mm)
1/64 — .015625		0.3969	33/64 — .515625		13.0969
1/32 — .03125		0.7938	17/32 — .53125		13.4938
3/64 — .046875		1.1906	35/64 — .546875		13.8906
1/16 — .0625		1.5875	9/16 — .5625		14.2875
5/64 — .078125		1.9844	37/64 — .578125		14.6844
3/32 — .09375		2.3813	19/32 — .59375		15.0813
7/64 — .109375		2.7781	39/64 — .609375		15.4781
1/8 — .1250		3.1750	5/8 — .6250		15.8750
9/64 — .140625		3.5719	41/64 — .640625		16.2719
5/32 — .15625		3.9688	21/32 — .65625		16.6688
11/64 — .171875		4.3656	43/64 — .671875		17.0656
3/16 — .1875		4.7625	11/16 — .6875		17.4625
13/64 — .203125		5.1594	45/64 — .703125		17.8594
7/32 — .21875		5.5563	23/32 — .71875		18.2563
15/64 — .234375		5.9531	47/64 — .734375		18.6531
1/4 — .250		6.3500	3/4 — .750		19.0500
17/64 — .265625		6.7469	49/64 — .765625		19.4469
9/32 — .28125		7.1438	25/32 — .78125		19.8438
19/64 — .296875		7.5406	51/64 — .796875		20.2406
5/16 — .3125		7.9375	13/16 — .8125		20.6375
21/64 — .328125		8.3384	53/64 — .828125		21.0344
11/32 — .34375		8.7313	27/32 — .84375		21.4313
23/64 — .359375		9.1281	55/64 — .859375		21.8281
3/8 — .3750		9.5250	7/8 — .8750		22.2250
25/64 — .390625		9.9219	57/64 — .890625		22.6219
13/32 — .40625		10.3188	29/32 — .90625		23.0188
27/64 — .421875		10.7156	59/64 — .921875		23.4156
7/16 — .4375		11.1125	15/16 — .9375		23.8125
29/64 — .453125		11.5094	61/64 — .953125		24.2094
15/32 — .46875		11.9063	31/32 — .96875		24.6063
31/64 — .484375		12.3031	63/64 — .984375		25.0031
1/2 — .500		12.7000	1 — 1.000		25.4000

APPENDIX B

CIRCUMFERENCES AND AREAS (0.2 to 9.8; 10 to 99)*

Diameter	Circum.	Area	Diameter	Circum.	Area	Diameter	Circum.	Area
0.2	0.628	0.0314	11	34.55	95.03	56	175.9	2,463
0.4	1.26	0.1256	12	37.69	113	57	179.1	2,551.8
0.6	1.88	0.2827	13	40.84	132.7	58	182.2	2,642.1
0.8	2.51	0.5026	14	43.98	153.9	59	185.4	2,734
1	3.14	0.7854	15	47.12	176.7	60	188.5	2,827.4
1.2	3.77	1.131	16	50.26	201	61	191.6	2,922.5
1.4	4.39	1.539	17	53.4	226.9	62	194.8	3,019.1
1.6	5.02	2.011	18	56.54	254.4	63	197.9	3,117.3
1.8	5.65	2.545	19	59.69	283.5	64	201.1	3,217
2	6.28	3.142	20	62.83	314.1	65	204.2	3,318.3
2.2	6.91	3.801	21	65.97	346.3	66	207.3	3,421.2
2.4	7.53	4.524	22	69.11	380.1	67	210.5	3,525.7
2.6	8.16	5.309	23	72.25	415.4	68	213.6	3,631.7
2.8	8.79	6.158	24	75.39	452.3	69	216.8	3,739.3
3	9.42	7.069	25	78.54	490.8	70	219.9	3,848.5
3.2	10.05	7.548	26	81.68	530.9	71	223.1	3,959.2
3.4	10.68	8.553	27	84.82	572.5	72	226.2	4,071.5
3.6	11.3	10.18	28	87.96	615.7	73	229.3	4,185.4
3.8	11.93	11.34	29	91.1	660.5	74	232.5	4,300.8
4	12.57	12.57	30	94.24	706.8	75	235.6	4,417.9
4.2	13.19	13.85	31	97.39	754.8	76	238.8	4,536.5
4.4	13.82	15.21	32	100.5	804.2	77	241.9	4,656.6
4.6	14.45	16.62	33	103.7	855.3	78	245	4,778.4
4.8	15.08	18.1	34	106.8	907.9	79	248.2	4,901.7
5	15.7	19.63	35	110	962.1	80	251.3	5,026.6
5.2	16.33	21.24	36	113.1	1,017.9	81	254.5	5,153
5.4	16.96	22.9	37	116.2	1,075.2	82	257.6	5,281
5.6	17.59	24.63	38	119.4	1,134.1	83	260.8	5,410.6
5.8	18.22	26.42	39	122.5	1,194.6	84	263.9	5,541.8
6	18.84	28.27	40	125.7	1,256.6	85	267.0	5,674.5
6.2	19.47	30.19	41	128.8	1,320.3	86	270.2	5,808.8
6.4	20.1	32.17	42	131.9	1,385.4	87	273.3	5,944.7
6.6	20.73	34.21	43	135.1	1,452.2	88	276.5	6,082.1
6.8	21.36	36.32	44	138.2	1,520.5	89	279.6	6,221.2
7	21.99	38.48	45	141.4	1,590.4	90	282.7	6,361.7
7.2	22.61	40.72	46	144.5	1,661.9	91	285.9	6,503.9
7.4	23.24	43.01	47	147.7	1,734.9	92	289.0	6,647.6
7.6	23.87	45.36	48	150.8	1,809.6	93	292.2	6,792.9
7.8	24.5	47.78	49	153.9	1,885.7	94	295.2	6,939.8
8	25.13	50.27	50	157.1	1,963.5	95	298.5	7,088.2
8.2	25.76	52.81	51	160.2	2,042.8	96	301.6	7,238.2
8.4	26.38	55.42	52	163.4	2,123.7	97	304.7	7,389.8
8.6	27.01	58.09	53	166.5	2,206.2	98	307.9	7,543.0
8.8	27.64	60.82	54	169.6	2,290.2	99	311.9	7,697.7
9	28.27	63.62	55	172.8	2,375.8			
9.2	28.9	66.48						
9.4	29.53	69.4						
9.6	30.15	72.38						
9.8	30.78	75.43						
10	31.41	78.54						

*The formulas for circumference and area of circles are the same regardless of the system of measurement, so these values are accurate for both inches and millimetres.

APPENDIX C

PROPERTIES, GRADE NUMBERS & USAGES			
Class of Steel	***Grade Number**	**Properties**	**Uses**
Carbon - Mild 0.3% carbon	10xx	Tough - Less Strength	Rivets - Hooks - Chains - Shafts - Pressed Steel Products
Carbon - Medium 0.3% to 0.6% carbon	10xx	Tough & Strong	Gears - Shafts - Studs - Various Machine Parts
Carbon - Hard 1.6% to 1.7%	10xx	Less Tough Much Harder	Drills - Knives - Saws
Nickel	20xx	Tough & Strong	Axles - Connecting Rods - Crank Shafts
Nickel Chromium	30xx	Tough & Strong	Rings Gears - Shafts - Piston Pins - Bolts - Studs - Screws
Molybdenum	40xx	Very Strong	Forgings - Shafts - Gears - Cams
Chromium	50xx	Hard W/Strength & Toughness	Ball Bearings - Roller Bearing - Springs - Gears - Shafts
Chromium Vanadium	60xx	Hard & Strong	Shafts - Axles -Gears - Dies - Punches - Drills
Chromium Nickel Stainless	60xx	Rust Resistance	Food Containers - Medical/Dental Surgical Instruments
Silicon - Manganese	90xx	Springiness	Large Springs

*The first two numbers indicate type of steel, the last two numbers indicate the approx. average carbon content — 1010 steel indicates, carbon steel w/approx. 0.10% carbon.

APPENDIX D

U.S. STANDARD GAUGES OF SHEET METAL

GAUGE	THICKNESS		WT. PER SQ. FT.		GAUGE
10	.1406″	3.571 MM	5.625 LBS	2.551 Kg.	10
11	.1250″	3.175 MM	5.000 LBS	2.267 Kg.	11
12	.1094″	2.778 MM	4.375 LBS	1.984 Kg.	12
13	.0938″	2.383 MM	3.750 LBS	1.700 Kg.	13
14	.0781″	1.983 MM	3.125 LBS	1.417 Kg.	14
15	.0703″	1.786 MM	2.813 LBS	1.276 Kg.	15
16	.0625″	1.588 MM	2.510 LBS	1.134 Kg.	16
17	.0563″	1.430 MM	2.250 LBS	1.021 Kg.	17
18	.0500″	1.270 MM	2.000 LBS	0.907 Kg.	18
19	.0438″	1.111 MM	1.750 LBS	0.794 Kg.	19
20	.0375″	0.953 MM	1.500 LBS	0.680 Kg.	20
21	.0344″	0.877 MM	1.375 LBS	0.624 Kg.	21
22	.0313″	0.795 MM	1.250 LBS	0.567 Kg.	22
23	.0280″	0.714 MM	1.125 LBS	0.510 Kg.	23
24	.0250″	0.635 MM	1.000 LBS	0.454 Kg.	24
25	.0219″	0.556 MM	0.875 LBS	0.397 Kg.	25
26	.0188″	0.478 MM	0.750 LBS	0.340 Kg.	26
27	.0172″	0.437 MM	0.687 LBS	0.312 Kg.	27
28	.0156″	0.396 MM	0.625 LBS	0.283 Kg.	28
29	.0141″	0.358 MM	0.563 LBS	0.255 Kg.	29
30	.0120″	0.318 MM	0.500 LBS	0.227 Kg.	30

WEIGHTS OF MATERIALS

Material	Avg. Lbs. per Cu. Ft.	Avg. Kg. per Cu. Metre	Material	Avg. Lbs. per Cu. Ft.	Avg. Kg. per Cu. Metre
Aluminum	167.1	2676	Mahogany, Honduras, dry	35	564
Brass, cast	519	8296	Manganese	465	7448
Brass, rolled	527	8437	Masonry, granite or		
Brick, common and			limestone	165	2648
hard	125	2012	Nickel, rolled	541	8649
Bronze, copper 8, tin 1	546	8754	Oak, live, perfectly dry		
Cement, Portland, 376 lbs.			.88 to 1.02	59.3	953
net per bbl	110–115	1765–1836	Pine, white, perfectly dry	25	388
Concrete, conglomerate,			Pine, yellow, southern dry	45	706
with Portland cement	150	2400	Plastics, molded	74–137	1200–2187
Copper, cast	542	8684	Rubber, manufactured	95	1518
Copper, rolled	555	8896	Slate, granulated	95	1518
Fibre, hard	87	1377	Snow, freshly fallen	5–15	70–247
Fir, Douglas	31	494	Spruce, dry	29	459
Glass, window or plate	162	2577	Steel	489.6	7837
Gravel, round	100–125	1586–2012	Tin, cast	459	7342
Iron, cast	450	7201	Walnut, black, perfectly dry	38	600
Iron, wrought	480	7695	Water, distilled or pure rain	62.4	988
Lead, commercial	710	11,367	Zinc or spelter, cast	443	7095

INDEX

604.2   c 1
NELSON
    How to read and
understand blueprints.

South St. Paul Public Library
106 Third Avenue N.
South St. Paul, MN 55075

DEMCO